Physical-chemical Properties of Foods

Modeling and Control of Food Processes Set

coordinated by
Jack Legrand and Gilles Trystram

Physical-chemical Properties of Foods

New Tools for Prediction

Aïchatou Musavu Ndob
Malik Melas
André Lebert

ELSEVIER

First published 2015 in Great Britain and the United States by ISTE Press Ltd and Elsevier Ltd

ISTE Press Ltd
27-37 St George's Road
London SW19 4EU
UK

www.iste.co.uk

Elsevier Ltd
The Boulevard, Langford Lane
Kidlington, Oxford, OX5 1GB
UK

www.elsevier.com

Notices

For information on all our publications visit our website at http://store.elsevier.com/

© ISTE Press Ltd 2015
The rights of Aïchatou Musavu Ndob, Malik Melas and André Lebert to be identified as the authors of this work have been asserted by them in accordance with the Copyright, Designs and Patents Act 1988.

British Library Cataloguing-in-Publication Data
A CIP record for this book is available from the British Library
Library of Congress Cataloging in Publication Data
A catalog record for this book is available from the Library of Congress
ISBN 978-1-78548-007-2

Printed and bound in the UK and US

Contents

Introduction

A physical-chemical property of a food corresponds to a particular measure of the state of the food at a given time and at a given place [RAH 09]. The understanding of physical-chemical properties is essential for engineers and scientists to help them solve problems of processing, of preservation, of storage, of distribution and even of food consumption. In a general manner, conservation and transformation processes alter the physical-chemical properties in a positive and a negative way. During food processing, the means of action to obtain defined properties of use can be gathered into three groups:

– control of the characteristics of the food item by adding ingredients (preservatives) or by removing elements detrimental to quality;

– use of energy in different forms (heat, light, etc.);

– control or elimination of any (re)contamination.

Physical-chemical properties of biological media such as pH, water activity (a_w) and the redox potential (E_h) have – along with temperature – a central role in the fields of biotechnology, pharmaceutical industries and food industries. As a matter of fact, they allow the description and the prediction of both the behavior of microorganisms and chemical and biochemical kinetics occurring in liquid, gelled or solid biological media. These media contain as a major component water – the solvent – and a number of solutes that may be neutral molecules such as sugars or alcohols or even electrolytes similar to salts, acids, bases or amino acids. Physical-chemical properties depend on the different solutes present in the environment, as well as on the concentration of these solutes.

Numerous studies focus on the production of foods undergoing less processing and/or containing fewer additives or preservatives. Therefore, in recent years, French (Na⁻) or European programs (TERIFIQ) have been focusing on the reduction of the quantities of salt, sugar and fat in foods. If health benefits for consumers are obvious – hypertension, cardiovascular diseases, diabetes and obesity reduction – it becomes not only necessary to adapt or even to completely modify the current processing and preservation processes but also to verify that organoleptic, technological and health properties are preserved. The decrease in the salt levels, for example, in cooked ham is reflected, for instance, by less water retention, by a less efficient adhesion of the muscles in them and by an increased microbial risk. However, the measurement of the evolution of these physical-chemical properties depending on the composition and the conditions of the process is often costly in time and money.

This book aims to show that it is possible to model and predict some of the physical-chemical properties (pH, a_w and ionic strength) for more or less complex liquid or solid media containing various solutes. The existence of such a model is very important. Actually, it makes it possible to meet the demands of engineers in the fields of biotechnology and pharmaceutical and food industries to lay out equipment and imagine new processes. As a matter of fact, a large amount of data about the equilibrium properties of media is needed to describe the processing of raw materials into finished goods or to develop new products with given characteristics. However, the low amount of available data does not currently allow such demands to be met. This explains the benefit of this modeling method that is part of a more general approach to design new products and/or new processes by numerical simulation.

1

The Main Physical-chemical Properties

Simulation processes and their optimization require knowledge of the physical-chemical properties of the pure substances and the mixtures being implemented [TOU 14]. This constitutes the field of study of applied thermodynamics that allows, from the molecular representation of pure substances and mixtures, the prediction of the physical-chemical properties while ensuring consistency with experimental data.

1.1. Physical-chemical properties and qualities of biological products

The purpose of this section is not to make a comprehensive survey of all the knowledge of the relationships between the physical-chemical properties and the qualities of food products, but merely to show by means of examples the need to measure and predict these properties.

Physical-chemical properties of biological media (pH, a_w and E_h) are crucial for the preservation of food products, for their nutritional qualities to be maintained and for the development of their organoleptic properties. Physical-chemical properties are related to the composition of the food (salt concentration, water, organic acid amounts, etc.): any modification in food composition or in the process causes a change of the physical-chemical properties, and therefore of the final product.

Water activity is a property of equilibrium, it is constantly used to evaluate the diffusion of water during processing. The relationship between the amount adsorbed (equivalent to the water content) of various gases such as oxygen, nitrogen and water vapor depending on their activity coefficient (equivalent to water activity) was put forward by Brunauer, Emmett and Teller at the end of the 1930s, then gradually extended to water and biological products. This connection is often represented in the form of sorption isotherms [RAH 09]. This explains why the concept of water activity is used by both researchers and technologists. Therefore, it is necessary to understand several additional aspects such as the methods of measurement of water activity and sorption curves, the models of representation and prediction of water activity and the curves of sorption as well as the relationships between water activity and microbial growth or chemical and enzymatic reactions.

Microorganisms require water, nutrients, a suitable temperature and an acceptable level of pH for their growth. Table 1.1 lists the acceptable range of values of pH for the growth of various microorganisms. The dissociated form of an organic acid is the form that has an inhibitory action on the microorganisms. This explains that weak acids (i.e. propionic acid, sorbic acid, etc.) are used to inhibit bacterial growth. Thus, at a pH = 5, 35% of acetic acids ($pK_a = 4.76$) are in an undissociated form versus 6% of lactic acid ($pK_a = 3.86$) and 0.4% of citric acid ($pK_{a1} = 3.13$, $pK_{a2} = 4.76$ and $pK_{a3} = 6.40$): for this reason, it should be wiser to make a mayonnaise with vinegar instead of lemon in order to limit health hazards. As a result, knowledge of the pK_a values is essential to explain or envisage, for example, the absence or not of inhibition according to the acid being used. The pH also has an effect on the production of toxins [BAB 14], but often in conjunction with the a_w [OH 14].

Organism	Minimum pH	Optimum pH	Maximum pH
Thiobacillus thiooxidans	0.5	2.0–2.8	4.0–6.0
Sulfolobus acidocaldarius	1.0	2.0–3.0	5.0
Bacillus acidocaldarius	2.0	4.0	6.0
Zymomonas lindneri	3.5	5.5–6.0	7.5
Lactobacillus acidophilus	4.0–4.6	5.8–6.6	6.8
Staphylococcus aureus	4.2	7.0–7.5	9.3
Escherichia coli	4.4	6.0–7.0	9.0
Clostridium sporogenes	5.0–5.8	6.0–7.6	8.5–9.0
Erwinia caratovora	5.6	7.1	9.3
Pseudomonas aeruginosa	5.6	6.6–7.0	8.0
Thiobacillus novellus	5.7	7.0	9.0
Streptococcus pneumonia	6.5	7.8	8.3
Nitrobacter sp	6.6	7.6–8.6	10.0

Table 1.1. *Minimum pH, optimum pH and maximum pH growth values of several bacteria (source: [RAH 07])*

In general, if the main interest lies in the processing procedures, it is necessary to simultaneously take into account several physical-chemical properties. This is, for example, the case of cheese, where a change in the nature or in the concentration of brine during manufacture causes an alteration in the concentration of salts in the cheese, which locally modifies the physical-chemical properties. Thus, the pH affects the functional properties and the texture of the cheese [PAS 03, GUI 02]. Similarly, water activity is a control and command factor of the organoleptic qualities or biochemical evolution [SHE 09]. These two physical-chemical qualities are also important selection mechanisms of microorganisms and, subsequently, of different floras during the manufacturing and ripening of the cheeses. This phenomenon is not limited to health protection with regard to pathogenic or alteration microorganisms, it also concerns the microflorae that show a technological relevance such as lactic acid and propionic bacteria.

In the case of meat products such as dried ham, salt plays an important role in the development of flavors as well as in the texture of the product. As a matter of fact, the pH, the concentration of NaCl and the temperature are the factors that most influence proteolysis during ham curing. An ultimate low pH ($<$ 5.7) promotes the release of proteases, whereas a strong concentration in NaCl is an inhibitor for the activity of proteases. As a result, the decrease in NaCl in dry ham not only causes problems of health safety but also texture defects (flabby and doughy) [HAR 14].

The role of redox potential (E_h) is much less known, particularly due to the difficulty of measuring it. However, it intervenes in the oxidation reactions of lipids and proteins, in the selection of microorganisms: aerobic microorganisms that grow when E_h is positive or anaerobic when E_h is negative.

1.2. Semi-empirical modeling of physical-chemical properties

In this section, the empirical equations that allow the sorption curves to be represented are not described. Rahman [RAH 09] conducts an exhaustive census of these equations and the parameter values for a relatively large number of food products. However, several semi-empirical approaches will be described because they can predict the pH or a_w of complex media.

1.2.1. *The Rougier et al. model (2007)*

Rougier *et al.* [ROU 07] have developed a predictive model of the a_w of food media rich in animal protein (such as fish, pork and beef meats), salted or high in fats. Their works have been carried out with gelatin to which salt and/or fats were added. They have measured the sorption curves using the Baucour and Daudin method [BAU 00]. The obtained values of a_w have then been adjusted with the Ferro Fontan model [FER 82]:

$$\ln\left(\frac{\gamma}{a_w}\right) = \alpha . X_W^{-R} \tag{1.1}$$

with γ, α and R adjusted parameters:

– X_W: water content in gelatin (kg water/kg dry matter).

After adjusting all the sorption curves, Rougier *et al.* [ROU 07] developed a model of the evolution of the parameters *γ*, *α* and R depending on the composition in salt, fats and proteins. As a result, they accounted for the interactions between water, proteins and lipids. However, since lipid globules do not interact with the protein matrix, their presence simply leads to an increase in the amount of dry matter [IGL 77]. The equation of the curves of sorption for lipid-rich gelatin is thus written as:

$$X_{W\,cal}^{DM} = \left(\frac{\ln(\gamma)/a_w}{x}\right)^{-(1/R)} \times (1 - X_{fat}^{DM}) \tag{1.2}$$

or even:

$$a_{w\,cal} = \gamma \exp\left(-\alpha \left(\frac{X_W^{DM}}{1 - X_{fat}^{DM}}\right)^{-R}\right) \tag{1.3}$$

with *γ*, *α* and R adjusted parameters of the Ferro Fontan equation:

– X_W: water content of the gelatin (kg water/kg dry matter);

– X_{fat}: lipid content of the gelatin (kg lipid/kg dry matter);

– DM = dry matter.

Figure 1.1 shows the comparison between measured curves and predicted curves. The concordance is quite satisfactory.

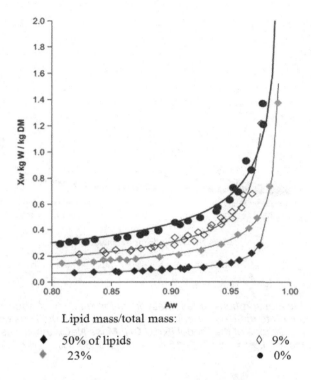

Figure 1.1. *Impact of lipid content on sorption isotherms at 20°C on gelatin gels. The points correspond to the experimental measurements and the lines correspond to the model predictions [ROU 07]*

In order to develop the prediction model of gelatin sorption curves with regard to the amount of sodium chloride, it is necessary to take account of the possible crystallization of this salt. Chrirife and Resnik [CHI 84] showed that from 0.351 kg_{Nacl}/kg_{eau}, a solution of sodium chloride is saturated: the dissolved species coexists with the crystallized salt. The a_w of the salted gelatin is calculated from the Ferro Fontan parameters (pure gelatin sorption curve) and a regression of the experimental data of Chirife and Resnik [CHI 84] for a solution of NaCl at 20°C. The detailed equations can be found in Rougier *et al.* [ROU 07]. Gelatin sorption curves are predicted by integrating an adjustment parameter (Figure 1.2).

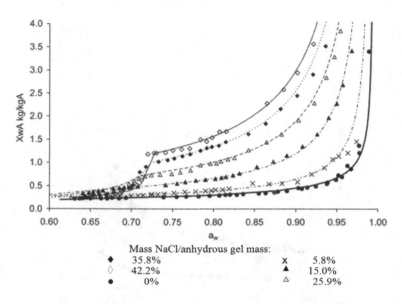

Figure 1.2. *Gelatin sorption curves relatively to the quantity of added salt. The points correspond to the experimental measurements and the lines correspond to the predictions of the model [ROU 07]. Mass NaCl/anhydrous gel mass: • 0%; x 5.8% ; ▲15% ; △ 25.9%; 35.8%; ◊ 42.2%*

This model allows forecasts for high moisture values of a_w ($a_w > 0.6$) at 20°C for a gelatin gel containing 0–50% w/w of lipid or 0–45% w/w of NaCl. Rougier *et al.* [ROU 07] finally used their model with success to predict the sorption isotherms of foods rich in protein such as cod [DOE 82], pork [COM 00] and beef [TRU 03].

A first limitation of this model lies in the difficulty of taking into account the crystallization of NaCl. Another limitation, substantially more significant, appears if the objective is to include a new ingredient or to replace all or part of the NaCl by KCl or $CaCl_2$. This actually implies on the one hand repetition a fairly large number of measurements of sorption curves and on the other hand a refit of the model.

1.2.2. *The a_w designer*

In order to help the food industries to predict the fate of their products, while limiting the number of experimental tests, ADRIA Développement offers a software program for the calculation of the water activity, the main concept of which is to calculate and predict the water activity of a formulation based on a database of over 230 ingredients and different models for liquid media [TEN 81] and the Guggenheim Anderson Boer (GAB) [RAH 97] and Ferro Fontan models [FER 82] for solid media.

The a_w designer – a paid-for tool – makes it possible to calculate the a_w of dairy, sea, meat, vegetable and cereal products as well as sauces and ready meals. It can help us to optimize the formulations to obtain microbiological stability, determine and compare the depressant activities of ingredients and reformulate the products. Nonetheless, as in the model of Rougier *et al.* [ROU 07], the addition of a new product or a new additive can only be carried out through new measures and further adjustments of parameters.

1.2.3. *The Wilson et al. model (2000)*

Foods and other biological systems are complex buffer systems to which weak acids are added as preservative agents. These weak acids (acetic acid, lactic acid, sorbic acid, etc.) allow the prevention or control of microbial growth in these numerous systems. Wilson *et al.* [WIL 00] have developed a model to predict alterations of pH in complex buffer systems resulting from the addition of weak acids. The theory describing the behavior of weak acids is well established. It requires knowledge of the constants of dissociation (pK_a) and the concentrations. However, many systems encountered in food or biological products are complex, and therefore poorly defined (microorganism culture media or meat and dairy products) and have several acids and weak bases. The nature of these molecules is rarely known, and their close proximity further complicates the understanding of their behavior (dissociation, etc.). The buffering capacity of these complex systems is often characterized empirically by the quantity of strong acid needed to change the pH, for instance, between a pH of 6 and 4 [MUC 91]. Wilson *et al.* [WIL 00] use a procedure with successive iteration to estimate the ratio between undissociated and dissociated forms as well as the pH. The basic equation is Henderson–Hasselbalch's. This method can be used to predict changes in pH caused by polyacids such as tricarboxylic acids, considering each acid function as a separate acid defined by its own pK_a. Figure 1.3 shows the results of this model for the titration of a brain-heart infusion (BHI) culture medium by four weak acids (acetic, lactic, formic and citric). Errors about the expected pH are on average less than 0.1 pH unit (upH) with a maximum of 0.2 upH.

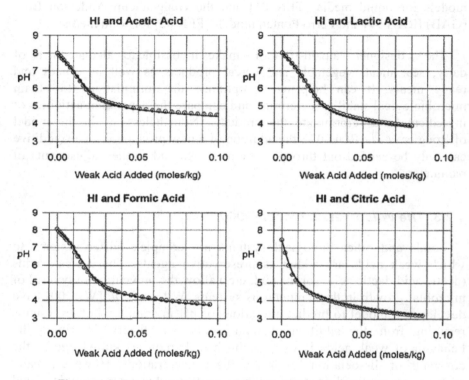

Figure 1.3. *Titration curves by weak acids with brain-heart infusion broth. Comparison with the predictions of the model of Wilson et al. [WIL 00]. The points correspond to the experimental measurements and the lines correspond to the predictions of the model*

A large number of food media contain a lipid phase and an aqueous phase in which microorganisms can grow. This situation occurs frequently in food systems such as sauces, for which weak acids are used to ensure food safety. A modified form of the Henderson–Hasselbalch equation has been developed taking into account the partitioning of the undissociated form of the acid in an oil phase, in order to calculate the proportion of weak acid present in its undissociated form in the aqueous phase.

However, to operate this model requires knowledge of the pK_a. Nonetheless, many dissociation constants remain unmeasured (for example, for more than half of the dipeptides) as well as the equilibrium constants between two phases.

1.3. From thermodynamic state quantities to the physical-chemical properties of food

Due to the limits of the above-mentioned approaches, it is interesting to address the problem of prediction of physical-chemical properties from a different angle. The proposed approach is based on thermodynamics. A methodology has been developed from the exact definition of the physical-chemical properties. This methodology allows us to overcome much of the previous limits.

1.3.1. *Quick thermodynamics primer*

The thermodynamic properties of a system (pure substance or mixture) are numerous, but experiments have shown that a limited number of these quantities suffice for the knowledge of a system and for the determination of all the other properties. These quantities are called state quantities, all the others being state functions (that is functions of state quantities) which by nature are extensive[1]. The number of state variables is characteristic of the studied system, the choice of state variables among the state quantities being largely arbitrary.

In order to describe the properties of a homogeneous fluid, three state variables are necessary among the following quantities: temperature (T), pressure (P), volume (V) and number of moles (n), the fourth being a state quantity. There is a relationship between these four quantities which can be the ideal gas equation or Van der Waals gas equation.

---

1 In thermodynamics, an extensive quantity characterizing a physical system is proportional to the size of this system, the system being assumed at equilibrium and homogeneous. During the reunion of two systems, the value of the quantity is equal to the sum of the values of this quantity for the two disjointed systems.

Classical thermodynamics defines several state functions:

– the internal energy (U) whose variation during a process is equal to the sum of heat and work transfers between the system and the external medium;

– the enthalpy that is the total energy of the thermodynamic system; it is the sum of the internal energy and the work that this system should exercise against the outside pressure to fill its volume;

$$H = U + P.V \qquad\qquad [1.4]$$

– the entropy (S) that characterizes the disorder of a system and that allows us to quantitatively identify the second law of thermodynamics;

– the free enthalpy or Gibbs energy (G) including:

- the growth during an isothermal and isobaric process represents the minimal value of the useful work that must be provided to the system to perform the process,

- the decrease represents the maximal value of the useful work that an operator can expect to collect during this process;

$$G = H - T.S \qquad\qquad [1.5]$$

– the free energy or Helmholtz energy (F or A) whose variation allows us to obtain the useful work that can be provided by a closed thermodynamic system, at a constant temperature, during a reversible process.

$$F = U - T.S \qquad\qquad [1.6]$$

Given an extensive quantity E^2 of a mixture of p components. In the case of a homogeneous phase, E is a function of temperature, pressure and the number of moles of each of the p constituents. The total differential of E is then written as:

$$dE = \left(\frac{\partial E}{\partial T}\right)_{P,n} + \left(\frac{\partial E}{\partial P}\right)_{T,n} + \Sigma_{i=1}^{1=p} dn_i.e_i \qquad\qquad [1.7]$$

---

2 The quantity E may be the volume, internal energy, enthalpy, entropy, Gibbs energy or Helmotz energy.

with:

$$e_i = \left(\frac{\partial E}{\partial n_i}\right)_{T,P,n_{j\neq i}} \qquad [1.8]$$

The quantity e_i is called partial molar quantity. Since E is an extensive quantity, then the variables E and e_i are connected by the relation:

$$E = \sum_{i=1}^{1=p} n_i. e_i \qquad [1.9]$$

By differentiating equation [1.9]:

$$dE = \sum_{i=1}^{1=p} dn_i. e_i + \sum_{i=1}^{1=p} n_i. de_i \qquad [1.10]$$

The combination of [1.7] and [1.10] allows the obtention of the Gibbs–Duhem equation:

$$\left(\frac{\partial E}{\partial T}\right)_{P,n} + \left(\frac{\partial E}{\partial P}\right)_{T,n} - \sum_{i=1}^{1=p} n_i. de_i = 0 \qquad [1.11]$$

This equation is simplified at constant temperature and pressure:

$$\sum_{i=1}^{1=p} n_i. de_i = 0 \qquad [1.12]$$

1.3.2. Chemical potential, activity and activity coefficient

Thermodynamics defines several partial molar quantities including the partial molar Gibbs energy which is also known as the chemical potential:

$$g_i = \mu_i = \left(\frac{\partial G}{\partial n_i}\right)_{T,P,n_{j\neq i}} \qquad [1.13]$$

The chemical potential depends on the composition of the mixture. It is defined as follows:

$$\mu_i = \mu_i^0 + R. T. \ln(a_i) \qquad [1.14]$$

Thus, the chemical potential depends on two variables: the free enthalpy of formation μ_i^0 in the chosen reference state and of activity a_i (which also depends on the reference state).

The chemical potential of a constituent is a measure of the ability of the constituent i to cause a physical or chemical transformation in a solution: a constituent with a high chemical potential shows great ability to develop a reaction or another physical process. It also translates the contribution of constituent i to the Gibbs energy of the mixture [BOT 91]:

$$G = \sum_i n_i \mu_i \qquad [1.15]$$

The activity of a constituent i is defined as the deviation to the selected reference. It can be expressed in different scales of concentration:

– in the scale of molar fractions (mainly used in the chemical industry), the activity is defined by:

$$a_i = \gamma_i x_i \qquad [1.16]$$

where γ_i is the activity coefficient:

– x_i is the mole fraction;

– the reference scale is at infinite dilution in water;

– in the scale of molalities[3] (mostly used for aqueous solutions and/or electrolyte solutions), the activity is defined by:

$$a_i = \gamma_i^m m_i \qquad [1.17]$$

where γ_i^m is the activity coefficient:

– m_i is the molality of constituent i in mol/kg water;

– the reference scale is at infinite dilution in water.

In biological media, which are real liquid solutions, the deviation from the reference state and thus the deviation to the ideal behavior is generally large because of the interactions between molecules. The ideal solution is only a theoretical concept which is being approximated when the solution is very diluted, in other words, in a condition close to the infinite dilution [ACH 92a]. Consequently, the characterization of the chemical potentials of the solution is based on the choice of a reference state for each of the constituents of the system [ACH 92b].

---

3 The molality of a solute is defined as the number of moles of the solute for a kilogram of solvent.

1.3.3. *Activities and physical-chemical properties*

Equation [1.14] giving the chemical potential of a constituent i may first be applied to water:

$$a_i = a_{H_2O} = a_w \qquad\qquad [1.18]$$

Similarly, the pH corresponds (in absolute value) to the activity of the hydrated proton H_3O^+:

$$pH = -\log(a_{H_3O^+}) = -\log(\gamma^m_{H_3O^+} m_{H_3O^+}) \qquad\qquad [1.19]$$

As mH_3O^+ is very similar to the concentration (expressed in mol/water liter) for dilute solution, a simplifier expression is obtained in which $\gamma^m_{H_3O^+}$ is equal to 1. This corresponds to an ideality assumption of the solution, or more precisely to the application domain of Henry's law. This approximation is only valid for dilute solutions and this dilution hypothesis must concern all the solutes available:

$$pH = -\log(C_{H_3O^+}) \qquad\qquad [1.20]$$

For a strong acid solution, this relation is only verified when the solutions are not very concentrated. For $c_0 = 0.1$ mol/L, the deviations between the ideal pH and real pH are already significant. For example, the conventional calculation of pH by [1.20] for a solution of hydrochloric acid of concentration $c_0 = 0.1$ mol/L would give a pH of 1.0. In fact, the real value of the pH is 1.16 and actually corresponds to $-\log(a_{H_3O^+})$ [BOT 91]. Experimentally, it is actually the activity of the ions H_3O^+ that is measured with a voltmeter placed at the terminals of an electrochemical cell.

Similarly, the pH can be altered by the addition of a solute: the Gibbs–Duhem law tells us that new chemical equilibria occur, such as dissociation or complexation equilibria, the proportion of each form depending on the single or several dissociation constants of the acid or the base. Therefore, a buffer solution KH_2PO_4/K_2HPO_4 of 0.1 M initially at pH 7 undergoes a change in pH during the addition of NaCl: the pH decreases to 6.4 with the addition of one mole of NaCl per kg of water (Figure 1.4).

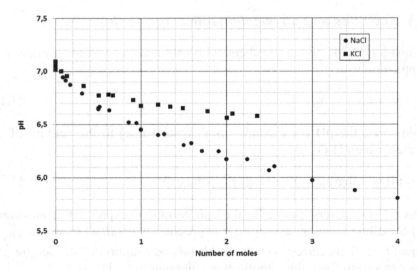

Figure 1.4. *Evolution of the pH in a buffer solution KH_2PO_4/K_2HPO_4 0.1 M depending on the number of moles of NaCl or Kcl addes [DES 04]*

In 1923, Joannes Brönsted and Thomas Lowry, independently, defined an acid as a species that tends to lose a proton, and a base as a species that tends to obtain a proton. Thus, for each acid AH, there is a conjugate base A^- and, for each base B, a conjugate acid BH^+. A strong acid AH (or BH^+) has a great tendency to lose a proton which indicates that the conjugate base A^- (or B) is a weak base that only has a low tendency to accept the proton. In an aqueous solution, the reaction is written as:

$$AH \rightleftharpoons H^+ + A^- \text{ or } H+ \rightleftharpoons H + +B \qquad [1.21]$$

Or even taking into account that the proton is hydrated:

$$AH + H_2O \rightleftharpoons H_3O^+ + A^- \text{ or } BH+_+H2O \rightleftharpoons H^3O + +B \qquad [1.22]$$

In a dilute aqueous solution, water being the solvent, its activity may be considered to be equal to 1. The dissociation constant is then written as:

$$K_a = \frac{a_{H^+} \cdot a_B}{a_A} = \frac{a_{H_3O^+} \cdot a_B}{a_A} \qquad [1.23]$$

Hence, the relation:

$$pK_a = pH + \log\left(\frac{a_A}{a_B}\right) \tag{1.24}$$

These various considerations about the pH and pK_a enable three important points to be made:

– a chemical equilibrium can be shifted by the only addition of a salt chemical unrelated to the species being considered. As such, there is no indifferent ion, because they are all involved in fixing the ionic strength;

– the experimental determinations of equilibrium constants should be conducted in diluted media if we want to get closer to the best of the conditions that allow the assimilation of concentrations and activities;

– it is by extrapolation to infinite dilution of experimental results that safe and reliable values of the equilibrium constants can be obtained.

In the case of a redox reaction, it is necessary to know the stoichiometry and the electrostatic potential of the medium (in volts) to determine the properties of equilibrium of the reaction:

$$ox + n\,e^- \rightleftharpoons red \tag{1.25}$$

By definition, the chemical potential of the electron is bound to the redox potential (E) by the relation:

$$\mu_{e^-} = -\mathcal{F}.E \tag{1.26}$$

where $\mathcal{F}$ is the Faraday constant $\mathcal{F} = 96\,485\ C.^{mol-1}$.

Ould Moulaye [MOU 98] showed that the calculation of the redox potential is equivalent to the calculation of the chemical potential of the electron. As a result, the Gibbs free enthalpy of reaction was equal to:

$$\Delta\mu = \mu_{red} - \mu_{ox} - n\,\mu_{e^-} \tag{1.27}$$

At the thermodynamic equilibrium ($\Delta\mu = 0$) and using [1.14], the chemical potential of the electron is equal to:

$$\mu_{e^-} = \frac{1}{n}\left(\mu_{red} - \mu_{ox}\right)$$

$$= \frac{1}{n}\left(\left(\mu_{red}^0 + R.T.\ln(a_{red})\right) - \left(\mu_{ox}^0 + R.T.\ln(a_{ox})\right)\right) \qquad [1.28]$$

The value of μ_{e^-} corresponds to the standard potential of the ox/red redox couple, denoted by E^0, defined in reference conditions, that is at a temperature of 25°C, and an ideal solution in which the oxidant and reductant are each at a reference molality of 1 mol.kg^{-1}. Thus, the standard potential is equal to:

$$E^0 = -\frac{\mu_{e^-}}{\mathcal{F}} = \frac{1}{n\mathcal{F}}\left(\mu_{red}^0 - \mu_{ox}^0\right) \qquad [1.29]$$

The redox potential of the solution is finally written as:

$$E_h = E^0 + \frac{R.T}{n\mathcal{F}}\ln\left(\frac{a_{ox}}{a_{red}}\right) \qquad [1.30]$$

So, the prediction of the a_w, pH and E_h is equivalent to predicting water, proton and electron activities. In the particular case of E_h, it is also necessary to know the formation properties. Consequently, it is necessary to resort to:

– thermodynamics for the prediction of the pH and water activity;

– thermodynamics and thermochemistry to predict the redox potential of a solution and the pKa of an acid/base couple.

Thermodynamics makes it possible to take into account the deviation from the ideality of the biological media. In the case of media containing ions, these corrections should be taken into account starting from weak concentrations: as mentioned previously, the addition of a salt can significantly alter the chemical equilibrium. In the case of media containing non-electrolytes, corrections of activity become significant when the concentration of a solute increases or when the number of solutes increases even if each one has a weak concentration. Figure 1.5 shows all of the relationships between physical-chemical parameters that can be calculated from the activity coefficient and the standard potential.

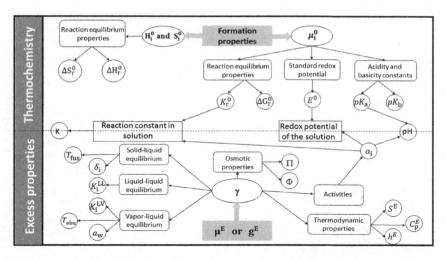

Figure 1.5. *The relationships between activity coefficient, standard redox potential and physical-chemical properties [TOU 14]*

Figure 1.7 The relationship between ... coefficient on values of ...
temperature and electrolyte concentration (3.5, 15)

A Thermodynamic Approach for Predicting Physical-chemical Properties

2.1. A brief historical overview

Since the beginning of physical chemistry, tens of thousands of articles have been written in order to understand the behavior of fluid mixtures. Although there is no general theory on the properties of fluid mixtures, a large number of theories and models whose area of application is restricted to a particular type of mixture are available [PRA 99]. The Gibbs free energy is a function of the mole fractions of the different constituents of the mixture. A first approach consists therefore of writing a limited expansion of this function and, according to the order of the expansion, the Margules, Redlich–Kister or Van Laar equations appear [HAL 68, PRA 99]. However, although series expansion is very flexible, (1) it does not explain physical and chemical phenomena occurring in a mixture and (2) it has mainly no predictive power since the phase transition of binary mixtures to ternary mixtures requires the adjustment of additional parameters.

In order to construct a theory of liquid mixtures, it is necessary to know two types of information: the structure of liquids (i.e. how the molecules in the liquid are disposed in space) and the intermolecular forces between similar or dissimilar molecules. However, information about the second type of information is insufficient and as a consequence all the theories must make simplifying hypotheses to overcome this drawback [PRA 99]. Theoretical works have concerned liquid mixtures whose molecules are apolar and spherical in shape: for example, the theory independently developed in 1933 by Scatchard and Hildebrand [PRA 99] frequently used for hydrocarbon mixtures by virtue of the quality of these predictions. All these theories have subsequently been extended – in a semi-empirical way in general – to more complexly shaped molecules. Thus Wilson [WIL 64] has developed a model in which the notion of local composition is introduced: the concentration of molecules i around a molecule j is different from the concentration of molecules j around a molecule i. Renon and Prausnitz [REN 68] introduced this concept into Scott's equation [SCO 56] and obtained the non-random two liquid (NRTL) equation. This equation constitutes an advance insofar as, based on the parameters of binary mixtures, it becomes possible to predict the behavior of ternary mixtures without any additional parameters to adjust.

With the advent of computer resources at the end of the 1960s, the prediction methods of the activity coefficient that have been developed rely both on the concept of local composition as well as on the principle of group contribution. In the group contribution method, the basic idea is that if tens of millions of different chemical compounds have been identified, the number of functional groups constituting these compounds is much narrower: depending on models, this number varies between 30 and 100. Expanding the idea of group contribution to mixtures is thus very attractive: while the number of pure compounds may be high and consequently the number of possible mixtures much more significant by several orders of magnitude, the number of functional groups remains the same. The problem is therefore reduced to the representation of a low number of functional groups and of their interactions. The first fundamental assumption of the group contribution method is to assume that the physical property of a fluid is the sum of contributions due to the functional groups of molecules. This makes it possible to correlate the properties of a large number of fluids based on a small number of parameters that characterize individually the contributions of the groups. The second fundamental hypothesis of a group contribution method is additivity: the contribution of a group in a molecule is independent of that of another group in this molecule. This hypothesis is valid only when the influence of a group in a molecule is not affected by the nature of the other groups within this molecule. Consequently, any one group contribution method is necessarily approximate since the contribution of a particular group in a molecule is not necessarily the same as in another molecule.

Several methods of group contributions have been developed: Analytical Solution of Groups (ASOG) [WIL 62], Universal Quasichemical

(UNIQUAC) [ABR 75] and Uniquac Functional Group Activity Coefficient (UNIFAC) [FRE 75]. The UNIQUAC method requires the adjustment of two parameters from the measurements of liquid–vapor equilibrium in binary mixtures, in contrast for higher order mixtures (ternary, quaternary, etc.) it becomes predictive. The ASOG and UNIFAC methods are predictive: the parameters of these methods are identified once and for all: it is no longer necessary to perform experiments in order to calculate the mixture properties of new molecules. These groups' contribution methods are used every day in the chemical industry, new developments achieved by UNIFAC [WEI 87, LAR 87, GME 98] or ASOG [KOJ 79, TOG 90] in order to extend the application domain of these methods as well as their precision. New methods such as COSMOS-RS [ECK 02, KLA 95], group Contribution Model Solvation [LIN 99] and the Segment Contribution Solvation model [LIN 02], although validated for a number of cases, are still under development.

Models that can be applied to electrolytes are essentially based on the works (1) of Debye and Hückel [DEB 23], which from a simplified description of long-range interactions such as ion–ion have given a satisfactory prediction of salt activities for low concentrations, and (2) of Pitzer [PIT 73] that by generalizing the Debye–Hückel equation has allowed the prediction of salt activities up to saturation. Chen *et al.* [CHE 82] have extended the concept of local composition developed in the NRTL model [REN 68] to electrolyte solutions by introducing two important parameters: the local electroneutrality and the repulsion between ions of the same charge. Kikic *et al.* [KIL 91], in order to take into account the effects of salts on the vapor-liquid equilibrium, have added a Debye–Hückel term to the UNIFAC equation. However, in all cases, Robinson and Stokes [ROB 59], then Achard *et al.* [ACH 92b] and Ben Gaïda *et al.* [BEN 10] have clearly highlighted the need to define the ion entity in a solution in terms of hydration degree.

2.2. The structure of the thermodynamic model

In the field of chemical engineering, a very large number of models have been developed for non-ideal solutions. In this chapter, only the model developed by Achard [ACH 92a] will be presented as it provides the basis for the present work.

2.2.1. *The interactions to be taken into account*

Biological, liquid or solid media are complex because of the number of constituents present (sugars, lipids, carbohydrates, salts, organic acids, etc.) which leads to a wide variety of interactions between the species in solution. Three types of main interactions are identified [ACH 92a]:

– Short-range interactions resulting from dispersion forces and from differences in size and in form between molecules. When molecules of a different nature are mixed, their medium is no longer the same as in pure solutions. Molecules are in perpetual movement and by coming into contact, induce modifications of their energy potential. These intermolecular or Van der Waals forces cause deviations from ideality. The other component concerns the geometry (area, volume and shape) of mixed molecules. There is a mixing entropy contribution because there is a transition from an ordered state to a more disordered state.

– Interactions of chemical nature (short-range interactions) come from the chemical properties of groups present in molecules. These properties are of two kinds: they result from association phenomena with the establishment of hydrogen bonds between polar groups (OH, SH and NH) when the separation distance between molecules is moderate and from solvation phenomena. These latter become significant in aqueous solutions because water is a highly polar molecule that is involved in the solvation of ions, in hydrogen bonds with sugars, alcohols, etc.

– Interactions of electrostatic nature (long-range interactions) are due to the electrical charges of the ionic species after dissociation of the electrolytes. On the basis of these charged particles existing in the medium, electrochemical repulsion or attraction phenomena occur. These interactions occur regardless of the distance separating the two charged particles.

These energy interactions are grouped into two types, short-range interactions (SR, short-range) constituted by the chemical interactions and the interactions related to the size and to the shape of the molecules and by long-range interactions (LR, long-range), due to electrostatic interactions [ACH 92a]. Long-range forces dominate for dilute electrolyte solutions. Short-range forces become significant at high concentrations in electrolytes or in non-electrolytes, which gives solutions strongly not ideal because then the three types of interactions take place.

2.2.2. The UNIFAC model

The UNIFAC model originally developed by Fredenslund *et al.* [FRE 75] divides the activity coefficient into two parts:

$$\ln \ (\gamma_i) = \ln(\gamma_i^C) + \ln(\gamma_i^R) \qquad\qquad [2.1]$$

where γ_i^C is the combinational part of the activity coefficient of the compound i

γ_i^R is residual part of the activity coefficient of the compound i.

The combinational part takes into account the influence of the size and of the shape of the molecules on the non-ideality of the solutions. Each functional group is characterized by two structure parameters, one in relation to the volume of Van der Waals (r_i) and the other with the surface of Van der Waals (q_i). The residual part describes the contribution to the non-ideality of the energy interactions between molecules. These interactions are characterized by two parameters (a_{mn} and a_{nm}) determined by fitting to experimental data that concerns the physical properties related to the activity. The experimental data are often phase equilibrium data containing hydrocarbons, water, alcohol, ketones, amines, esters, ethers, aldehydes, chlorides, nitriles and other organic constituents. The UNIFAC model is fully predictive. The parameters are determined by fitting experimental data concerning a large number of very different constituents containing these groups [SIM 02]. From this knowledge base, the set of parameters is used to predict both the non-ideality of constituents not included in the base and the non-ideality of complex solutions (number of constituents > 3).

The original UNIFAC model has shown limitations notably about the prediction of the coefficients of infinite dilution activity, the predicted level of non-ideality was too significant. Moreover, this model was not correctly taking into account the impact of temperature in the calculations. As a result many changes have been made by modifying the combinational part and by calculating the binary interaction parameters between groups depending on the temperature [ACH 92a].

A particular model, the Dortmund modified UNIFAC model [SKJ 79], is still being revised and its application domain extended. Data from the literature and new experimental data are stored in the Dortmund data bank. This version allows us a better description of the effect of temperature and of the actual behavior in the dilute region; the model applies more properly for systems with molecules of different sizes. New functional groups are regularly added [GME 82, HAN 91, GME 98, LOH 01], but they essentially concern the fields of chemical separation processes.

Similar changes were performed for the UNIFAC model modified by [LAR 87], but its performance is not as good as the Dortmund model according to a comparative study conducted by Lohmann et al. [LOH 01]. However, this model is relevant because it properly estimates the activity coefficients of solutions containing sugars [ACH 92a]. Peres and Macedo [PER 97] have introduced in their model, two conformation groups of carbohydrates (pyranose and furanose) as well as an OH ring group taking into account the proximity of the OH groups in these molecules and a group –O– for disaccharides. They have recalculated the interaction parameters for D-glucose, D-fructose and sucrose. The model gives very accurate predictions of the a_w, of the boiling temperature and of the solubility for binary and ternary solutions and has been successfully applied to other carbohydrates and to carbohydrate solutions simulating juices [PER 99]. The prediction of the a_w has been correctly achieved on polyols such as glycerol, sorbitol or mannitol, which are solutes present in food, pharmaceutical or cosmetic fields [NIN 00]. Predictions of the Larsen modified UNIFAC model are sometimes not as good at high concentration, because the model does not differentiate the molecular structures of isomers. CH_2OH subgroups have been introduced by Spiliotis and Tassios [SPI 00] and take into account the position and the orientation of the hydroxyl group on each ring.

Other studies [OZD 99, NIN 00, GRO 03] show the relevance of UNIFAC models for the prediction of the a_w in agri-food industries or in biological systems. However, sugars or polyols are not the only solutes that can reduce the a_w of a medium, in the same manner that the a_w is not the only property providing stability to food products. Electrolytes and the pH are also involved.

Appendix 1 gives the equations for calculating the combinational part and the residual part of the activity coefficient for the UNIFAC equation modified by Larsen et al. [LAR 87]. The volume (r_i) and surface groups (q_i)

parameters are estimated from molecular data of pure constituents and are available for non-electrolyte constituents in Appendix 4. Finally, Appendix 5 provides the values of the interaction parameters between functional groups.

2.2.3. *The electrolyte model*

The most used model to calculate the activity coefficient of an electrolyte solution is the Pitzer model [PIT 73b] based on the calculation of the electrostatic interactions between ions according to the Debye–Hückel theory [ACH 92a]. From a structural point of view, the solution of a solute in water causes changes in the properties of the solute and of water. The ions become solvated by surrounding themselves with water molecules. The water supply is partially destroyed and opposite electrical charges attract one another [BOT 91]. From an electrostatic point of view, at the location of an ion i, all the other ions create an electrical potential that depends on the distribution and on the charge of these ions with regard to ion i [BOT 91]. The electrical energy derived there from can be calculated and used to determine the activity coefficients of the present species. The electrical potential allows the calculation of the ionic strength I which is defined by:

$$I = \frac{1}{2}\left[\sum_{i_+} c_{i_+} z_{i_+}^2 + \sum_{i_-} c_{i_-} z_{i_-}^2\right] \qquad [2.2]$$

where z_{i_+} and z_{i_-} are the charges of the ions, and

c_{i_+} and c_{i_-} are the molar concentrations in ions (mol/l of water).

From the activity coefficients of each dissociated species, it is also possible to calculate the mean activity of a CA formula electrolyte. If we consider the dissociation reaction of this electrolyte in cations and in anions, the reaction is written:

$$C_{v_C}A_{v_A} \rightleftharpoons v_C\, C^{z_C} + v_A\, A^{z_A}$$

where: v_C and v_A are the valence of ions;

v_A and z_A are their electrical charges.

For the CA electrolyte, the mean ionic molality $m_\pm$ and the average activity coefficient $\gamma_\pm$ [PRA 99] are defined as follows:

$$m_{\pm} = \left(m_C^{\nu_C} \, m_A^{\nu_A}\right)^{1/(\nu_C + \nu_A)} \tag{2.3}$$

$$\gamma_{\pm} = \left(\gamma_C^{\nu_C} \, \gamma_A^{\nu_A}\right)^{1/(\nu_C + \nu_A)} \tag{2.4}$$

A mean activity $ma_{\pm}$ is then obtained:

$$a_{\pm} = \left(a_C^{\nu_C} \, a_A^{\nu_A}\right)^{1/(\nu_C + \nu_A)} = m_{\pm} \cdot \gamma_{\pm} \tag{2.5}$$

The Pitzer model includes a modified Debye–Hückel-like contribution and a virial term to take short range interactions into account. Only two parameters having physical meaning must be adjusted. Accurate results have been obtained for the properties of electrolyte solutions attaining molalities up to 6 moles/kg of solvent. Activity coefficients have been calculated from this model for solutions containing different salts. They have been correctly predicted for solutions of NaCl, KCl and CaCl$_2$ (from [DEM 91]).

2.3. The Achard model

A more global approach consists of coupling models taking into account short-range interactions and models taking into account long-range interactions. Achard [ACH 92a] has conducted a review of the various existing models and then has developed its own model that we present in this section. It concerns a combined model that facilitates the calculation of activity coefficients of solutions containing electrolytes and uncharged molecules.

2.3.1. *The Achard model structure*

The Achard model combines the UNIFAC group contribution model modified by Larsen *et al.* [LAR 87], the Pitzer–Debye–Hückel equation [PIT 73a, PIT 73b] and solvation equations (Figure 2.1). The latter are based on the definition of the number of hydration for each ion, which corresponds to the assumed number of water molecules chemically related to the charged species. It divides the activity coefficient into two terms:

$$ln(\gamma_i) = ln\left(\gamma_i^{SR}\right) + ln\left(\gamma_i^{LR}\right) \tag{2.6}$$

γ_i^{SR} takes into account the short-range interactions (SR) (chemical forces, forces related to the size and to the shape of the molecules, solvation phenomena between water and the ionic species). γ_i^{LR} takes into account long-range interactions (LR) such as electrostatic ones. The detail of the equations is thoroughly presented in the different appendices.

The solution model thus built provides the detailed composition of the solution (concentration of the various charged species) and the activities of the different solutes. The activities calculated by the model can be directly compared with the experimental values of a_w and of the pH because it is a predictive model. The optimization of the interaction parameters of the model, already carried out on a database related to binary solutions, allows the model for ternary or more solutions to be used.

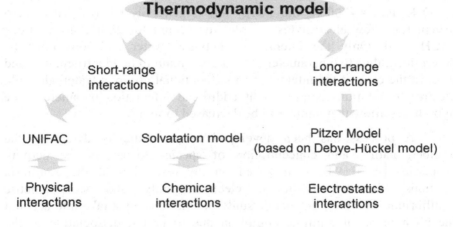

Figure 2.1. *Structure of the Achard model [ACH 92a]*

The implementation of this model has required that knowledge be inferred in three areas concerning the structural parameters, the interaction parameters and the detail of the species present in the solution (Figure 2.2):

– The charged ions and the groups' structural parameters R_k and Q_k were not available in the tables. Their estimate has been obtained from the crystal ionic radius and taking account of solvation phenomena, that is of the number of water molecules solvating the ion. Achard *et al.* [ACH 94] calculated the radius of an ion solvated by water molecules. They have made the assumption that solvation phenomena were constant, this is justified as long as the electrolyte concentration is not too significant.

– The interaction parameters between ions and water or between ions and the other functional groups of molecules have been adjusted in the Larsen modified UNIFAC model. These adjustments have been made from experimental data. Because of the limited amount of experimental data with regard to the coefficients to be identified, the interaction parameters a_{ij} have been expressed according to the interaction energies u_{ij} and u_{jj} ($a_{ij} = u_{ij} - u_{jj}$). In the case of water/ion interactions, a number of assumptions have been established to reduce the number of interaction energies to identify. As a result, some have been fixed. If a water–salt solution is being considered, there are three species: water (w), cation (C) and anion (A). The values taken for the u_{ij} are $u_{CC} = u_{AA} = 2,500$ K, $u_{ww} = -700$ K; the energy u_{CA} is generally taken equal to zero except for a few electrolytes in order to account for high concentrations [ACH 92a]. Only the interactions between water and ions must be determined, thus two parameters out of six remain to be identified, $u_{C,w}$ and $u_{A,w}$. In the case of the interactions ion/functional group, a water–alcohol–electrolyte solution requires eight additional interaction parameters, the hypotheses allow that number to be decreased to four.

– A process has been developed to automatically determine the number, nature and concentration of species to access the activity properties [ACH 94]. It is based on the resolution of the equilibria relations, the hypothesis of electroneutrality and stoichiometric equilibrium. The thermodynamic study of solutions must take into account the phenomena of multiple equilibria due to partial dissociation of the constituents. For example, when Na_2HPO_4 is dissolved in water, it yields the species: Na^+, H^+, OH^-, PO_4^{3-}, HPO_4^{2-}, $H_2PO_4^-$ and H_3PO_4. The species concerned by these dissociations are carboxylic acids (organic, fat), amino acids, organic bases (nitrogenous bases), inorganic weak acids (H_3PO_4 and H_2CO_3), inorganic weak bases (NH_3) and all acid or weak-base salts. These new electrically charged species modify the physical-chemical characteristics, the ionic strength, and create electrostatic interactions that contribute to the non-ideality of the solution.

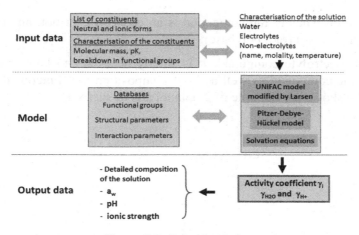

Figure 2.2. *Data structuring*

2.3.2. *Management of chemical species*

The Achard model includes a module for the automatic determination of the present chemical species. As a matter of fact, the decompositions and the solvations that can occur should be taken into account. Figure 2.3 shows two possible situations, one for a sugar and the other for a soda solution.

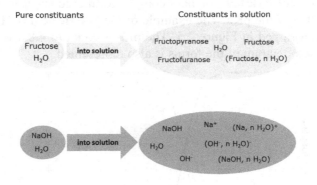

Figure 2.3. *Determination of the present chemical species*

2.3.3. *Decomposition into functional groups*

In order to operate, the Achard model requires a description of each chemical species present in solution (Figure 2.2). Several cases are possible depending on whether the compound is an electrolyte or not, an acid (or a base) or a polyacid (or a polybase):

– Non-electrolyte compound: it is simply necessary to know the molar mass of the compound as well as its decomposition into functional groups. Figure 2.4 shows the example of 2-butanol and of sucrose.

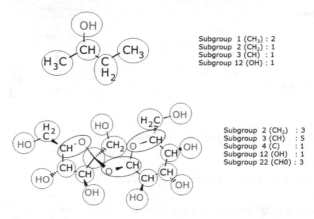

Subgroup 1 (CH₃) : 2
Subgroup 2 (CH₂) : 1
Subgroup 3 (CH) : 1
Subgroup 12 (OH) : 1

Subgroup 2 (CH₂) : 3
Subgroup 3 (CH) : 5
Subgroup 4 (C)   : 1
Subgroup 12 (OH) : 1
Subgroup 22 (CH0) : 3

Figure 2.4. *2-butanol and sucrose functional groups' decomposition*

– Acid-like (or base) or polyacid compound: moreover, it is necessary to know the value of the pKa. The example of the butyric acid (CH_3-CH_2-CH_2-COOH) is developed in Figure 2.5. The pKa is equal to 4.83. In this figure, the distribution diagram of the forms is also represented.

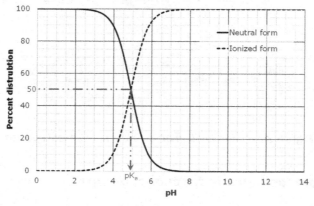

Figure 2.5. *Functional groups' decomposition of tartaric acid*

– Amino acid-like compound: when amino acids are present in aqueous solution and at a pH close to the isoelectric point (pI), they are present in the form of a zwitterion, a form bearing both a positive and a negative charge. Figure 2.6 clarifies the case of valine, as well as the domains of existence of each form depending on the pH. The valine pKa are equal to 2.3 and 9.7, and the pI is equal to 6.0.

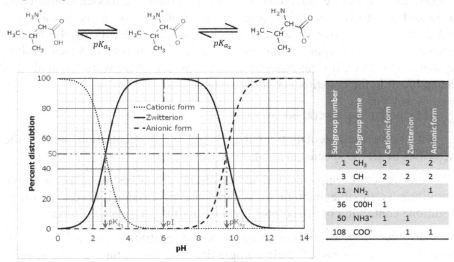

Figure 2.6. *Valine decomposition into functional groups*

– Salt-like compound: in aqueous solution, a salt is dissociated into two or more ions. Then, each ion binds either with OH⁻ or with H⁺ to yield a base or an acid. Thus, for sodium chloride:

$$NaCl \xrightarrow[in\ solution]{} Na^+ + Cl^- \tag{2.7}$$

then:

$$Na^+ + OH^- \rightarrow NaOH \tag{2.8}$$

and:

$$Cl^- + H^+ \rightarrow HCl \tag{2.9}$$

Finally, solvation phenomena of ions must be considered as it has been previously mentioned (Figure 2.3).

2.3.4. *Advantages and disadvantages of the Achard model*

Due to its structure, the Achard model is a predictive model. Considering a mixed electrolyte solution (e.g.: water–NaCl–KCl), it is possible to calculate the activities of all the species and constituents of the media, because the interaction parameters are from binary water-electrolyte systems. However, it is an aqueous solution model where water must always be included in the constituents of the system.

This model has several advantages:

– it allows the calculation of the pH, of the concentration, and of the activities of all the species in solution;

– the structure implemented to represent the phenomena of multiple equilibria is flexible enough to achieve the processing of very complex solutions containing or not solutes involved in dissociation equilibria;

– all the solutions containing low concentration solutes can be considered, regardless of their complexity and the nature of the species present.

There are however limitations that are related to the quality and the quantity of experimental data used for the adjustment of the parameters:

– the interaction energies which are not available were set to zero;

– the concentrations of gaseous constituent NH_3 and CO_2 are assumed to be very low and the interaction parameters a_{ij} regarding these gases have been zeroed;

– for the $H_2PO_4^-$ group, obtained from the dissociation of a weak electrolyte, the interaction energies should have been adjusted considering dissociation reactions and consequently all the existing forms (HPO_4^{2-} and PO_4^{3-}). Data from the literature have been obtained at acid pH where the ion $H_2PO_4^-$ is preponderant and the other forms negligible. Thus, Achard [ACH 92a] has used the only structural parameters available, i.e. those of $H_2PO_4^-$. Subsequently, Desnier-Lebert [DES 04] has therefore optimized some of the parameters concerning the phosphates;

– for solutions containing significant acid or organic bases, amino acid (m > 1 mol/kg of water) concentrations, the interaction energies between ions and functional groups such as COOH and NH_2 could not be obtained due to the scarcity of experimental data.

Applications to Biological Media

3.1. Simple media

3.1.1. *Sugar solutions*

The Achard model [ACH 92b] facilitates the identification of sugars such as glucose and fructose (Table 3.1). However, it does not allow the α and β forms to be separated.

Molecules	Chemical formula	Number of functional groups				
		Alkane CH$_2$	Alkane CH	Alkane C	Alcohol OH	Ether CH-O-
Glucose	$C_6H_{12}O_6$	1	4	0	5	1
Fructose	$C_6H_{12}O_6$	2	2	1	5	1
Maltose	$C_{12}H_{22}O_{11}$	2	7	0	8	3
Sucrose	$C_{12}H_{22}O_{11}$	3	5	1	8	3
Glycerol	$C_3H_8O_3$	2	1	0	3	0

Table 3.1. *Sugars and glycerol decomposition into functional groups*

The prediction of the a_w of the water + glycerol or water + maltose solutions is good regardless of the tested molality, up to 5.5 moles of glycerol/kg of water or 2.8 moles of maltose/kg of water (Table 3.2). The relative deviation (E) remains on average below 1.5%, that is 0.007 a_w unit. However, for high molality values, it can reach, for example 3% at 11 moles of fructose/kg of water and 6% at 5.8 moles of sucrose/kg of water. For water + glucose + fructose ternary solutions, the predictions are quite correct since the mean deviation is 0.53%.

Solute	Molality range (mole/kg of water)	E % (n)
Glucose	0–5.5	0.68 (9)
Sucrose	0–5.8	1.16 (11)
Glycerol	0–5.4	0.32 (9)
Fructose	0–11.1	0.99 (11)
Maltose	0–2.8	0.19 (9)

Table 3.2. *Mean deviations (%) between the experimental a_w and predicted a_w for sugars and glycerol. In brackets: n, number of measurements for each solute (source: Desnier-Lebert [DES 04])*

For quaternary solutions, water + phosphate buffer + a solute (glucose, sucrose or glycerol), the predictions are correct and less than 0.01 a_w units for molalities less than 3 moles/kg of water for sucrose and glycerol and less than 0.005 a_w unit for a same molality in glucose. The predicted a_w are similar whether the 0.1 M phosphate buffer is at pH 7.0 or pH 5.8 (Figure 3.1).

In quaternary solutions, water + pH 5.6 sodium diacetate buffer + glucose, the experimental pH are correctly predicted by the model: the deviations remain below 0.1 pH unit. However, in a buffer phosphate at pH 5.8 or pH 7.0, a decrease in pH is observed to be all the stronger if the glucose concentration is high. This decrease is not taken into account by the model, the deviation between the measured and predicted pH reaches

0.44 upH (pH unity) in the pH 7.0 buffer and 0.53 upH for the pH 5.8 buffer at 3.8 mol of glucose/kg of water.

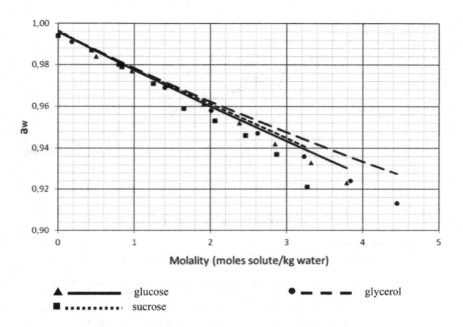

Figure 3.1. *Comparison between experimental and predicted a_w values in a 0.1 M KH_2PO_4/K_2HPO_4 buffer at pH 7 or pH 5.8 depending on the concentration of glucose, sucrose or glycerol. The points correspond to the experimental measurements and the lines correspond to the predictions of the model*

3.1.2. *Salt solutions*

The addition of KCl, KNO_2 or KNO_3 in KH_2PO_4/K_2HPO_4 pH 7 buffer causes a decrease in the pH, which is correctly predicted by the model. The pH is slightly underestimated by 0.1–0.2 upH irrespective of the studied concentration. During the addition of KCl in acetic acid/sodium acetate buffer at pH 5.6 or 0.1 M KH_2PO_4/K_2HPO_4 buffer at pH 6.2, the predictions are also correct. In the presence of NaCl, $NaNO_2$ or $NaNO_3$ (Figure 3.2), a decrease in pH is also observed and again predicted by the model regardless of the pH, nature of the sodium salts or tested molality: the deviations observed are below in average to 0.2 upH without exceeding 0.3 upH.

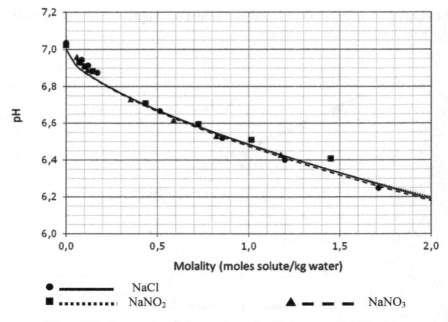

Figure 3.2. *Comparison between experimental and predicted values of the pH in a 0.1 M KH_2PO_4/K_2HPO_4 buffer depending on the salt concentration. The points correspond to the experimental measurements and the lines correspond to the predictions of the model*

3.1.3. *Amino acid solutions*

The UNIquac Functional group Activity Coefficient (UNIFAC) model modified by Larsen *et al.* [LAR 87] does not take into account the sulfur groups: among the common amino acids, methionine and cysteine cannot be decomposed into functional groups (Table 3.3).

From these decompositions, amino acid titration curves can be predicted. The evolution of pH is correctly predicted by the model for amino acids such as glycine, aspartic and L-glutamic acids (Figure 3.3). Most predictions for the other amino acids are correct, however, for some, small differences between measurements and predictions have been observed, because they are due to the pK_a values that are not generally measured at zero ionic strength and infinite dilution.

Amino acid	Formula	PM	CH₃	CH₂	CH	C	CHC	OH	CH₂CO	CH₂NH	CHNH	N	NH₂	NH₃⁺	ACH	AC	COOH	COO⁻
										Number of functional groups							
Alanine	$C_3H_7NO_2$	89.09	1		1									1				1
Arginine	$C_6H_{14}N_4O_2$	174.2		2	1					1		1	2	1				1
Asparagine	$C_4H_8N_2O_3$	132.12			1				1				1	1				1
Aspartic acid	$C_4H_7NO_4$	133.1		1	1									1			1	1
Cysteine	$C_3H_7NO_2S$	121.16	No possible decomposition															
Glycine	$C_2H_5NO_2$	75.07		1										1				1
Glutamine	$C_5H_{10}N_2O_3$	146.15	1	1					1				1	1				1
Glutamic acid	$C_5H_9NO_4$	147.13		2	1									1			1	1
Histidine	$C_6H_9N_3O_2$	155.16		1	1					1		1	1	1				1
Isoleucine	$C_6H_{13}NO_2$	131.17	2	1	2									1				1
Leucine	$C_6H_{13}NO_2$	131.17	2	1	2									1				1
Lysine	$C_6H_{14}N_2O_2$	146.19		4	1								1	1				1
Methionine	$C_5H_{11}NO_2S$	149.21	No possible decomposition															
Phenylalanine	$C_9H_{11}NO_2$	165.19		1	1									1	5	1		1
Proline	$C_5H_9NO_2$	115.13		3	1									1				1
Serine	$C_3H_7NO_3$	105.09	1	1				1						1				1
Threonine	$C_4H_9NO_3$	119.12	1		2			1						1				1
Tryptophan	$C_{11}H_{12}N_2O_2$	204.23	1		1	1					1			1	4	2		1
Tyrosine	$C_9H_{11}NO_3$	181.19		1	1			1						1	4	2		1
Valine	$C_5H_{11}NO_2$	117.15	2		2									1				1

Table 3.3. *Amino acids decomposition into functional groups*

Measurements were also made with mixtures of amino acids: they show that the thermodynamic model is applicable for the prediction of pH for ternary or quaternary solutions of amino acids (Figure 3.4).

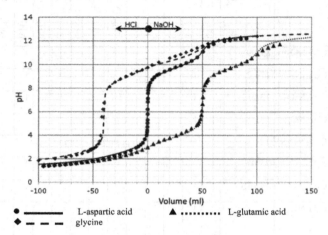

Figure 3.3. *Comparison of the experimental titration curves predicted by the model of three amino acids. The points correspond to the experimental measurements and the lines correspond to the predictions of the model*

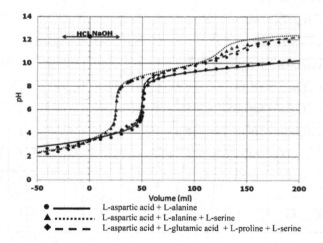

Figure 3.4. *Comparison of amino acid mixtures titration curves. The points correspond to the experimental measurements and the lines correspond to the predictions of the model*

3.2. Integration of complex media in the thermodynamic model

Taking into account certain components of complex media such as peptone, tryptone or meat extracts of microorganisms culture media can be achieved using three approaches:

1) Exhaustively describing all of the compounds present in the media. However, three major difficulties emerge:

– for biological media such as meats or derived from macromolecule hydrolysis (proteins, polyholosides, etc.), it is impossible to know all of the compounds;

– when the number of compounds is large, complexation phenomena, which are not taken into account by the thermodynamic model, may appear thus modifying the activities;

– for most of the compounds, several properties are unknown such as the pK_a: therefore, among the possible dipeptides, only the pK_a of approximately 10% have been measured.

2) Approximating the medium by its main constituents. Again, several difficulties are present:

– how to choose the constituents? Should only the most significant be chosen (in terms of mass or percentage)? The risk is not taking into account a constituent present in small amounts but having a strong influence, for example a salt;

– how many constituents should be taken? It is necessary that the number be large enough to reflect the composition of the medium but low enough to limit calculations. For example, if there are 50 constituents in the medium and 10 were chosen to be taken, there are $\binom{50}{10}$ combinations or 10,272,278,170 combinations to be tested.

3) Creating a fictional molecule with the same behavior as the medium. This molecule must predict the pH and a_w of the medium at different concentrations, in the presence of various additives such as acids, organic acids, salts, organic acid salts and sugars in order to simulate the evolution of the medium during operations such as drying or salting/brining and marinating.

It is this latter approach that has been implemented using the following methodology:

– creation of a database of titration curves and sorption curves of the medium studied in the presence or not of additives used in laboratories (for microorganisms culture media) or in the industry;

– creation of the fictional molecule:

- from the data of a single titration curve, determination of the molar mass and the number of pK_a,

- from the information available on the composition of the medium (acids, amino acids, lipids, carbohydrates, etc.), determination of the functional groups present in the medium,

- by combining the two previous pieces of information, determination of the number of each functional group by eliminating those whose number is less than 1,

- optimization of each pK_a value to best fit the predicted titration curve to the experimental titration curve;

– validation of the fictional molecule with the other titration curves and all the sorption curves. If results are not totally satisfactory, either the number of pK_a must be changed (increase in general) or the functional groups slightly modified.

This methodology has been successfully applied for microorganisms, gelatin and food products (meat and cheese) media cultures. The following sections only show the obtained key results with no detail, with the exception of the section related to the creation of the database.

3.3. Microorganisms culture media

Microorganisms culture media contain simple constituents (glucose, salt, etc.) and complex constituents deriving from the hydrolysis of proteins originating from animals or vegetables (peptones, tryptones, etc.) that are used as protein sources. Desnier-Lebert [DES 04] has used the Achard model to predict the evolution of the physical-chemical properties of the tryptone-meat broth medium (TMB) and the gelled tryptone-meat broth medium (TMBg). The TMB medium is composed of 10 g of proteose peptone/kg of water, 5 g of tryptone/kg of water, 10 g of meat extract/kg of water and 5 g of glucose per kg/water. The TMBg medium is composed of TMB medium and gelatin that allow for the solidification of the medium (Table 3.4).

	Ratio	Measured humidity (%)	Amount of added water	Amount of dry matter
Proteose peptone	2.50	34.00	0.85	1.65
Bacto tryptone	1.25	4.26	0.05	1.20
Meat extract	2.50	4.07	0.10	2.40
Glucose	1.25	0.11	0.00	1.25
Gelatin	120.00	12.00	14.40	105.60
Total	127.50		15.40	112.10

Table 3.4. *Ratio of each constituent in the TMBg. Distribution in water and dry material according to the humidity (kg of water/kg total matter) measured for each constituent*

3.3.1. Peptones and TMB medium

From a titration curve, the characteristics (molar mass and pK_a) of the three peptones used could be determined (Table 3.5). The decomposition into functional groups is carried out taking into account the following two constraints:

– decompose each peptone into three forms, acid, neutral and basic;

– have at least a NH_2 group and a COOH group, since they are composed of amino acids or peptides.

Peptone	Functional groups					Molar mass	pK_{a1}	pK_{a2}
	CH_3	CH_2	CH	NH_3^+	COO^-			
Proteose peptone	3	4	4	1	1	214.3	4.0	9.5
Tryptone	3	4	5	1	1	227.3	4.5	9.7
Meat extract	2	3	3	1	1	172.2	4.2	9.9

Table 3.5. *Decomposition into functional groups, molar mass and pK_a of the three peptones*

Since peptones are described as a constituent in the database of the thermodynamic software, they are used to estimate the pH and a_w of the TMB medium with a KH_2PO_4/K_2HPO_4 buffer at pH 5.8 or pH 7.0 and increasing concentrations of solutes (NaCl, KCl, glucose, sucrose and glycerol).

The prediction of the pH of the TMB is correct regardless of the salt or pH of the tested buffer (Figure 3.5). A deviation is nevertheless observed between the measured and calculated pH in the buffered TMB at pH 5.8 in the presence of NaCl. It reaches 0.2 pH unit at 2.6 moles of NaCl/kg of water.

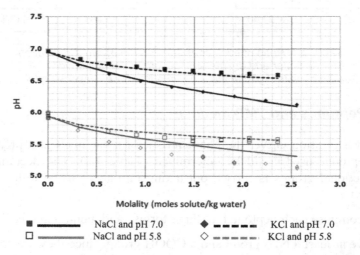

Figure 3.5. *Comparison between the experimental and predicted values of the pH of the TMB medium in a buffer 0.1 M KH₂PO₄/K₂HPO₄ depending on the salt concentration. The points correspond to the experimental measurements and the lines correspond to the predictions of the model*

However, the prediction of the a_w is very good regardless of the pH and the salt (Figure 3.6).

In the presence of sugars, some deviations of pH and a_w are observed starting from low molalities for the pH (0.2 upH for molalities below 1.5 moles solute/kg of water and 0.4 for molalities comprised between 1.5 and 4 moles solute/kg of water) and for the high molalities for the a_w (Figure 3.7). Satisfactory results were also obtained with glycerol.

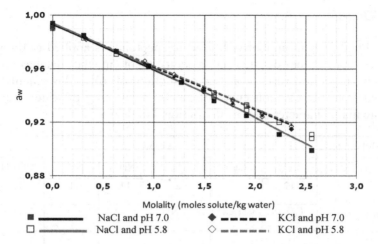

Figure 3.6. *Comparison between the experimental and predicted values of the a_w of the TMB media in a 0.1 M KH_2PO_4/K_2HPO_4 buffer depending on the concentration of salt. The points correspond to the experimental measurements and the lines correspond to the predictions of the model*

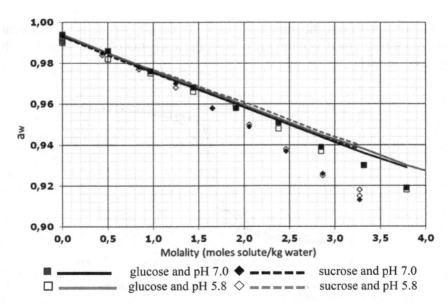

Figure 3.7. *Comparison between experimental and predicted values of a_w of the TMB medium in a 0.1 M KH_2PO_4/K_2HPO_4 buffer depending on the salt concentration. The points correspond to the experimental measurements and the lines correspond to the predictions of the model*

3.3.2. Gelatin

The implementation approach used for peptones is applied to the gelatin in order that it be defined as a constituent in the database of the thermodynamic software. For the estimation of pK_a, the calculation is achieved from the titration curve of gelatin at 50 g/kg of water. Table 3.6 gives the main characteristics of the fictional molecule.

Molar mass				1763.07		
Functional groups	CH$_3$	CH$_2$	CH	NH$_2$	NH$_3$$^+$	COOH
	12	14	12	11	1	11
	COO$^-$	CH$_2$CO	CH$_2$NH			
	1	7	7			
pK$_a$	1.2	2.2	3.9	8.4	10.9	12.9

Table 3.6. *Gelatin decomposition into functional groups, molar mass and pK$_a$*

Figure 3.8 shows the titration curves of gelatin solutions at different concentrations. The deviations are on average lower than 0.15 upH and reach at most 0.4 upH.

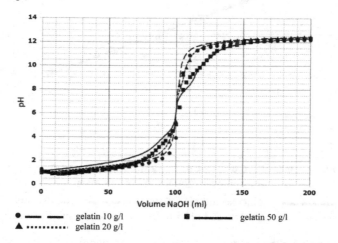

Figure 3.8. *Comparison of the experimental and predicted titration gelatin curves at different concentrations. The points correspond to the experimental measurements and the lines correspond to the predictions of the model*

The thermodynamic model also allows for the prediction of the pH of a gelatin solution in the presence of an acid and/or a salt (Figure 3.9). It should be noted that the model correctly predicts the lack of effect of the addition of salt on the pH.

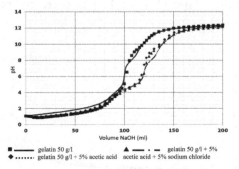

Figure 3.9. *Comparison of the experimental and predicted titration gelatin curves in the presence of sodium chloride acetic acid. The points correspond to the experimental measurements and the lines correspond to the predictions of the model*

However, so far all the tested media were liquid. A more important test for the thermodynamic model consists of not only checking the predictions of the a_w of a gelatin gel but also of the sorption curves according to the addition of salt. Up to a limit of 45% of added salt in a gelatin gel, the thermodynamic model allows the prediction of the sorption curves with a mean error of the order of 0.005 a_w (Figure 3.10). These results are significant because they show the ability of the thermodynamic model to be operational for gelled or even solid media.

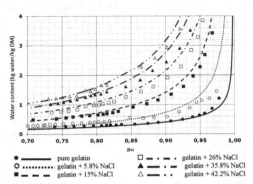

Figure 3.10. *Comparison of the experimental and predicted sorption curves for gelatin added or not to NaCl. The points correspond to the experimental measurements and the lines correspond to the predictions of the model*

3.3.3. TMBg medium

Figure 3.11 shows a good prediction of the sorption curves of the gelled TMBg medium in the range 0.75–1.00 for various organic acid salts. The influence of the cation type (Na^+ or K^+) is correctly predicted either for lactates or sorbates.

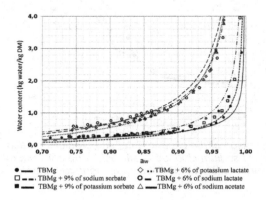

Figure 3.11. *Comparison of the predicted and experimental sorption curves for the TBMg medium with or without additives. The points correspond to the experimental measurements and the lines correspond to the predictions of the model*

3.4. Meat products

3.4.1. Beef and pork

In a similar way to the results showed by Rougier *et al.* [ROU 07], pork and beef can be represented by the same model. The characteristics of the fictional meat molecule have been calculated from a beef titration curve (Table 3.7).

Molar mass	\multicolumn{6}{c}{2229.58}					
	CH_3	CH_2	CH	NH_2	NH_3^+	COOH
	15	18	15	14	1	14
Functional groups	COO^-	CH_2CO	CH_2NH			
	1	9	9			
pK_a	2.5	3.5	3.9	7.0	9.0	11.6

Table 3.7. *Beef or pork decomposition into functional groups, molar mass and pK_a*

The predictions of beef and pork titration curves are very good. The deviations between experimental curves and predicted curves are on average lower than 0.2 upH in the pH domain ranging from 2 to 10. In the presence of additives (salts, organic acids and organic acid salts), the results show mean deviations of 0.3 upH with a minimum lower than 0.6 upH. Figure 3.12 shows the titration curves of beef in the presence of three organic acids: acetic acid, lactic acid and citric acid. Similarly to the case of gelatin, the model correctly predicts the absence of effect of adding a salt on the pH.

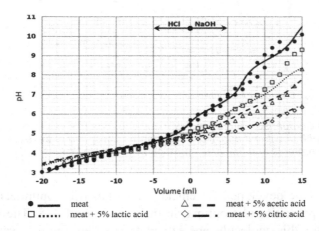

| | meat | △ — — | meat + 5% acetic acid |
| □ ...... | meat + 5% lactic acid | ◇ — . | meat + 5% citric acid |

Figure 3.12. *Comparison of the experimental and predicted titration curves of beef/pork in the presence of a sodium chloride organic acid. The points correspond to the experimental measurements and the lines correspond to the predictions of the model*

The sorption curves are also correctly predicted (Figure 3.13) in the presence of increasing amounts of sodium chloride. The deviations are less than 0.01 a_w unit when the a_w is less than 0.95 and comprised between 0.01 and 0.02 a_w unit for an a_w greater than 0.95.

3.4.2. *Poultry meat*

Comparable work has been carried out on chicken meat and has allowed the development of an equivalent molecule with a molar mass of 1,309.91 g/mole. The decomposition into functional groups was carried out from the average composition of chicken meat (Table 3.8), from the amino acid composition of the proteins and from the distribution of lipids in different fatty acids (Table 3.9).

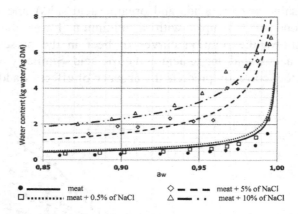

Figure 3.13. *Comparison of the predicted and experimental sorption curves for beef or pork with or without additives. The points correspond to the experimental measurements and the lines correspond to the predictions of the model*

The molecule developed for chicken meat has been optimized to ensure that the pH predictions be optimal in the pH range of 4–8. The predictions obtained with this molecule are satisfactory either for the water activity (Figure 3.14) or for the pH (Figure 3.15).

Average composition of chicken meat (g/100 g of muscle)	
Water	74.9
Total proteins	23.2
Total lipids	1.7
Total	99.8

Table 3.8. *Average composition of chicken meat*

Amino acid	Mass (g)	Fatty acid	Mass (g)
Alanine	1.266	Myristic acid	0.01
Arginine	1.399	Palmitic acid	0.280
Aspartic acid	2.068	Stearic acid	0.130
Glutamic acid	3.474	Palmitoleic acid	0.040
Cystine	0.297	Oleic acid	0.340
Glycine	1.140	Linoleic acid (LA)	0.220
Histidine	0.720	α-linolenic acid (ALA)	0.010
Isoleucine	1.225	Gadoleic acid	0.010
Leucine	1.741	Arachidonic acid (AA)	0.060
Lysine	1.971	Eicosapentaenoique acid (EPA)	0.010
Methionine	0.642	Clupanodonic acid (DPA)	0.010
Phenylalanine	0.921	Docosahexaenoic acid (DHA)	0.020
Proline	0.954		
Serine	0.798		
Threonine	0.980		
Tryptophan	0.271		
Tyrosine	0.783		
Valine	1.151		

Table 3.9. *Chicken meat composition in amino acids and fatty acids*

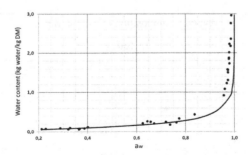

Figure 3.14. *Comparison of the experimental and predicted sorption curves for chicken meat. The points correspond to the experimental measurements and the lines correspond to the predictions of the model*

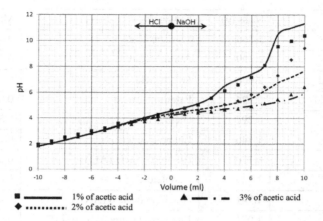

Figure 3.15. *Comparison of the experimental and predicted titration curves of chicken meat in the presence of acetic acid. The points correspond to the experimental measurements and the lines correspond to the predictions of the model*

3.4.3. *Fish*

The results presented in this section are preliminary. As a matter of fact, the first works carried out about fish flesh mainly concern cod and salmon. A problem, which at present is not yet totally resolved, concerns the high content of fat of the flesh of certain fish. This difficulty, highlighted by Iglesias and Chirife [IGL 77] and partly overcome by Rougier *et al.* [ROU 07], concerns the non-interaction between lipid globules and matrix protein, resulting in an increase in the content of dry matter. Table 3.10 gives the average composition of salmon flesh.

Average composition of salmon flesh (g/100g)	
Water	64.9
Total proteins	20.4
Ash	1.1
Total lipids	13.4
Total	99.8

Table 3.10. *Average composition of salmon flesh*

If the early results are quite satisfactory for the pH (Figure 3.16), they must be further improved with regard to the a_w. Several tracks are currently being followed to improve the results.

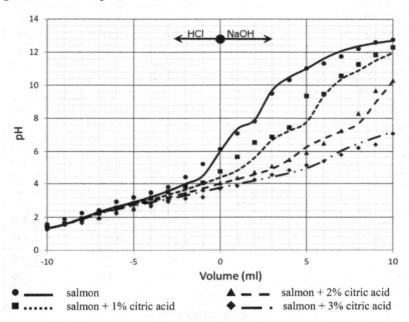

● ———	salmon	▲ — —	salmon + 2% citric acid
■ ······	salmon + 1% citric acid	◆ —— ·	salmon + 3% citric acid

Figure 3.16. *Comparison of the experimental and predicted titration curves of salmon flesh in the presence of citric acid. The points correspond to the experimental measurements and the lines correspond to the predictions of the model*

3.5. Dairy products

In 2003, Gros and Dussap used the initial Achard model to predict the physical-chemical properties of milk from the composition. They could thus find once more the equilibrium pH and the buffering capacity of the milk.

In recent works, four cheeses have been modeled to be incorporated into the thermodynamic model: fromage frais and three ripened cheese: Comté, Emmental and Edam. The nature and number of functional groups used are connected with the composition of part of the molecules of the cheeses (Table 3.11).

Compounds	Comté	Edam	Emmental	Fromage Frais
Amino acids				
Aspartic acid	2.171	1.747	1.569	0.905
Glutamic acid	7.895	6.150	5.704	2.603
Alanine	1.265	0.764	0.914	0.384
Arginine	1.283	0.964	0.927	0.497
Glycine	0.704	0.486	0.508	0.222
Histidine	1.475	1.034	1.065	0.326
Isoleucine	2.128	1.308	1.537	0.591
Leucine	4.095	2.570	2.959	1.116
Lysine	3.578	2.660	2.585	0.934
Phenylalanine	2.301	1.434	1.662	0.577
Proline	5.107	3.251	3.690	1.229
Serine	2.269	1.547	1.640	0.639
Threonine	1.437	0.932	1.038	0.500
Tryptophan	0.555	0.352	0.401	0.147
Tyrosine	2.344	1.457	1.693	0.604
Valine	2.960	1.810	2.139	0.748
Lipids				
Saturated fatty acids				
Caproic acid	0.9913	0.461	0.488	0.023
Caprylic acid	0.820	0.300	0.289	0.025
Capric acid	0.469	0.589	0.625	0.064
Lauric acid	1.203	0.496	0.521	0.070
Myristic acid	4.447	2.943	3.062	0.263
Unsaturated fatty acids				
Oleic acid	11.329	6.911	6.017	0.714
Palmitoleic acid	0.956	0.813	0.879	0.000
Linoleic acid	2.288	0.744	1.011	0.210
Carbohydrates				
Sucrose	0.100	0.000	0.000	0.140
Glucose	0.520	0.000	0.000	0.480

Table 3.11. *Comté, Emmental, Edam and fromage frais average composition (in g per 100 g of cheese)*

From the average compositions of each cheese and from the amino acids, the fatty acids and the sugars decomposition into functional groups, the functional groups composition of the studied cheeses is determined. Thus, the fictitious molecules of Edam and fromage frais are constituted of 15 groups, while those of Comté and Emmental are constituted of 12 groups with a preponderance of CH_2 groupings. Table 3.12 gives the characteristics of fromage frais.

Molar mass	2608.10					
Functional groups	CH_3	CH_2	CH	C	CH=CH	ACH
	7	31	13	1	1	4
	AC	OH	NH_2	NH_3^+	COOH	COO^-
	2	2	2	9	4	11
	CH_2NH	CHNH				
	1	1				
pK_a	1.5	2.0	3.7	5.9	9.7	11.6

Table 3.12. *Decomposition into functional groups, molar mass and pKa of the fromage frais*

The developed models have been tested to predict the cheeses titration and sorption curves in the presence of additives commonly used in the manufacturing or processing process of these foods (salts, acids and mixtures of salts and acids). The prediction range of the a_w is the same for the four cheeses (0.5–1.0), whereas the range of the pH is specific to each cheese: from 4 to 10 for the Comté, 3.5–12 for the Edam, 4–11 for the Emmental and 2–12 for the fromage frais.

The evolution of the pH differs slightly depending on the type of cheese: the pH of fromage frais varies quite significantly according to the addition of acid and base unlike the pH of the three other cheeses that remain substantially stable. When adding an organic acid, a shift in the titration curves can be observed toward acidic pH and is much more pronounced in the basic part of the titration curve. It depends on the nature and concentration of the organic acid: the increase in the concentration of the organic acid decreases the pH of the cheese with a more pronounced effect of the acetic acid. Finally, there is no modification in the titration curve when adding 5% of NaCl. The model correctly predicts the titration curves of the cheeses with or without the addition of various additives (Figures 3.17–3.19).

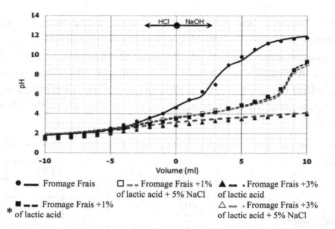

Figure 3.17. *Comparison of the experimental and predicted titration curves of the* fromage frais *in the presence of lactic acid and sodium chloride. The points correspond to the experimental measurements and the lines correspond to the predictions of the model*

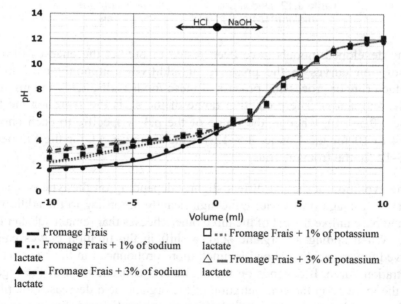

Figure 3.18. *Comparison of the experimental and predicted titration curves of fromage frais in the presence of sodium lactate or potassium lactate. The points correspond to the experimental measurements and the lines correspond to the predictions of the model*

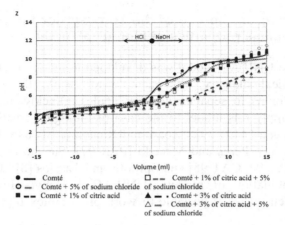

Figure 3.19. *Comparison of the experimental and predicted titration curves of the Comté in the presence of citric acid and sodium chloride. The points correspond to the experimental measurements and the lines correspond to the predictions of the model*

The prediction results of sorption curves by the thermodynamic approach are very satisfactory for the four cheeses: only the results for the fromage frais are presented as an example (Figure 3.20). The average deviations are less than 0.01 a_w unit and maxima deviations are less than 0.015 a_w unit in the a_w range 0.8–1.0.

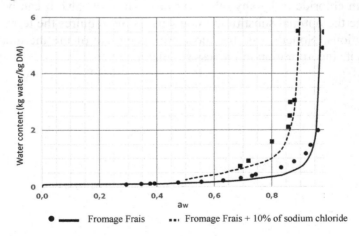

Figure 3.20. *Comparison of the experimental and predicted sorption curves in the presence of sodium chloride. The points correspond to the experimental measurements and the lines correspond to the predictions of the model*

3.6. Conclusion on the use of the software for the prediction of physical-chemical properties of biological media

Initially, the UNIFAC model was developed to meet the demands of the oil and chemical industry, for example to enable the calculation of distillation columns to separate new molecules. Quickly, tests have been conducted to apply this model to the case of biological media such as the works of Allaneau [ALL 79], Lebert and Richon [LEB 84a, LEB 84b], Le Maguer [LEM 92], Peres and Macedo [PER 97], Sancho et al. [SAN 97], Catté et al. [CAT 94, CAT 95] and Spiliotis and Tassios [SPI 00]. However, it is the combination of models (UNIFAC + Debye–Hückel Pitzer + solvation) which makes it possible to work on complex liquid, gelled and solid media.

The various examples that have just been presented show that it is possible, with limited experimental effort, to be able to predict the evolution of the physical-chemical properties of liquid biological media or solid foods. It then becomes easy to be able to quickly calculate the effect of composition changes on the pH or the a_w of the food, without going through tests expensive in time and money. Among the requests made by health authorities, the reduction of the quantities of sodium chloride has been included. The thermodynamics software allows us to quickly check the effect of such a reduction and to test the substitution of sodium chloride by potassium chloride or by any other salt or mixture of salts. It can be used to determine the optimal substitution rate which only requires the test of one or two solutions. Nonetheless, the most promising use of the thermodynamic model is its incorporation in process simulators.

Usage in Process Simulators

4.1. Introduction to numerical simulation

Numerical simulation is a study method that consists of analyzing a phenomenon, a process or the behavior of a system through a model [BIM 02]. While modeling consists of implementing the model, simulation makes use of the model that shows a behavior similar to the real system in order to consider a very large number of scenarios whose synthesis subsequently allows decision-making. During the design of new products, simulation must be regarded as an experiment, the term "software experiment" is sometimes used, that is as a tool for exploring possibilities. The use of computers facilitates the multiplication of studies, the enrichment of the cases studied and sometimes the correct break down of the factors having strong interactions between them.

However, is it really possible to create new products with numerical simulation? What is the necessary knowledge to positively respond to this question? In order to describe the transformation of raw agricultural products into more or less complex food (Figure 4.1), it is necessary to describe it in terms of time and space:

– heat and matter (water, salts, etc.) transfers that occur between the product and its environment or inside the product. Such transfers result in the emergence of temperature gradients, water content and solutes, and therefore by punctual modifications in the chemical composition of the product. They are characterized by exchange coefficients which depend on environmental conditions (temperature, relative humidity, etc.) and by diffusion coefficients which depend on physical-chemical conditions (temperature, water activity, pH, etc.);

– the chemical, biochemical or microbiological kinetics that occur on the surface or within the product. These kinetics also induce local modifications in the chemical composition;

– the local chemical composition modifications that are induced. These result in changes of the chemical potentials of the different chemical species present, and therefore of their activities (in the thermodynamic sense) as the pH and a_w. However, the kinetics velocity constants or the availability of the solutes actually depend on these activities;

– the evolution of mechanical, nutritional, hygienic and sensory properties of the product according to the transfers and kinetics. These properties are the result of the developments described previously.

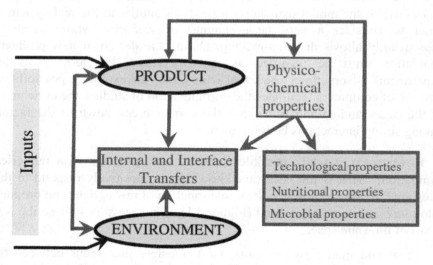

Figure 4.1. *Schematic representation of the relations between heat and materials and properties transfers of the products*

4.2. Prediction of the growth of micro-organisms during a drying process

4.2.1. *Presentation of the problem*

Lebert *et al.* [LEB 04, LEB 05] have studied the case of the growth/no growth of *Listeria innocua* on the surface of a gelatin gel undergoing drying. They have combined three types of models (Figure 4.2):

– heat and matter transfers models in order to obtain the evolution of the water content and the temperatures in the gelatin gel with regard to time and the drying conditions (temperature, velocity and relative humidity of the air);

– thermodynamic model that transforms the water content profiles in pH and a_w profiles;

– predictive microbiology model which from the growth conditions (temperature, pH and a_w) determines the latency and generation times of the bacterium.

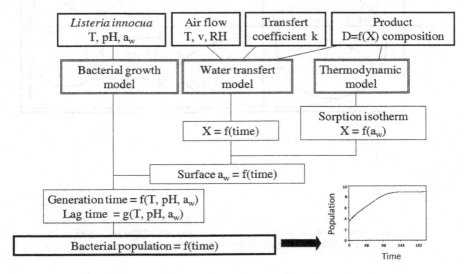

Figure 4.2. *Combination of models for the prediction of the growth of a bacterium according to time and the environmental conditions*

This global and comprehensive approach to the prediction of the quality and the microbiological safety of food, illustrated in Figure 4.3, requires resorting to several disciplines: microbiology, process engineering as well as physical-chemical or, in other terms, thermodynamic characterization.

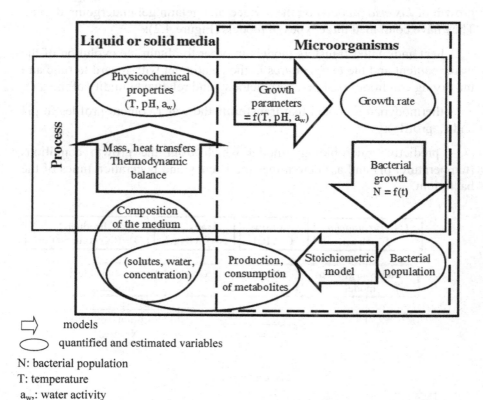

models
quantified and estimated variables

N: bacterial population
T: temperature
a_w: water activity

Figure 4.3. *Integrated modeling strategy of the physical-chemical properties of foodstuffs and the development of the bacterial populations undergoing a manufacturing process*

A food, liquid or solid medium, is characterized by physical-chemical properties – most often non-uniform in the product – that influence the development of micro-organisms. It must be described by growth (predictive microbiology) and metabolite production models: growth (latency time and generation time) parameters are estimated based on the variables that describe the microbial environment (temperature, pH and a_w). In this way, the growth of the bacterial population is estimated. This increase in population leads to the consumption and/or to the production of metabolites which can be represented by stoichiometric models taking into account the conversion of the carbon source into metabolism products of the bacterium. The compositions of the medium and the microbial environment are thus modified by bacteria activity as well as by water and solute transfers. The fact of taking into account the effect of the control variables of the processes on matter and heat transfers brings essential and additional information on the evolution of the composition of foods. This allows the calculation of the physical-chemical properties that in turn influence the microbial development. Such a scheme, therefore, brings forward complex interactions.

4.2.2. Growth curves

Growth tests of *L. innocua* on the surface of a gelatin gel have been carried out in a microbiological blower [ROB 98] at 18°C [LEB 05]. Figure 4.4 shows the evolution of the population in three different conditions:

– air velocity = 2.3 m/s and average relative air humidity = 95.5%;

– air velocity = 2.3 m/s and average relative air humidity = 91.9%;

– air velocity = 4.6 m/s and average relative air humidity = 92.6%.

as well as the result of the simulations. A good fit can be observed between measurements and predictions.

Nevertheless, the significance of this study does not lie in this good fit, but in the information given by the models – which is difficult to access or even impossible to measure – and that allows the results to be better understood.

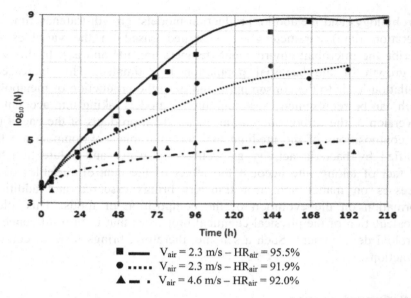

Figure 4.4. *Growth of Listeria innocua on the surface of a gelatin gel during drying at 18°C. The points correspond to the experimental measurements and the lines correspond to the predictions of the model*

4.2.3. *Use of the numerical simulation in order to explain growth curves*

Thus, if the most favorable condition is being considered (Figure 4.5), the relative humidity of the air has ranged from 94.5 to 96.5% throughout the test. The combination of the matter transfer model and the thermodynamic model allows the estimation of the water activity on the surface of the gelatin gel. The latter decreases at the beginning and then at the end of the test, whereas the relative humidity of the air rises, increasing slightly. The water activity remains above the growth limit of the *L. innocua* which had been determined between 0.935 and 0.945 [DES 04].

Now considering the most rigorous condition, that is an air velocity of 4.6 m/s and a relative air humidity of 92.6% (Figure 4.6). The relative air humidity does not fluctuate or very slightly around this average value. Consequently, the a_w on the gel surface decreases very quickly and as early as at 48 h it becomes less than 0.94, and therefore becomes less than the minimal growth a_w of the *L. innocua*. Thus, there is stunted growth.

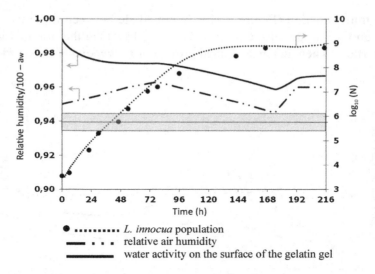

Figure 4.5. *Growth of L. innocua on the surface of a gelatin gel during drying at 18°C, 2.3 m/s and 95.5% HR. The points correspond to the experimental measurements and the lines correspond to the predictions of the model except for the relative humidity of the air which is measured*

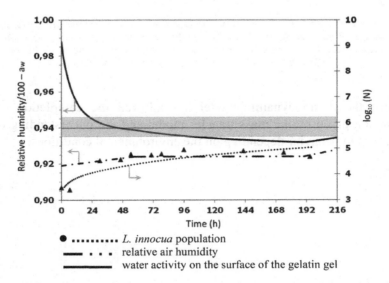

Figure 4.6. *Growth of L. innocua on the surface of a gelatin gel during drying at 18°C, 4.6 m/s and 92.6% HR. The points correspond to the experimental measurements and the lines correspond to the predictions of the model except for the relative humidity of the air which is measured*

The third situation (Figure 4.7) is intermediate between these two cases. The a_w on the surface of the product becomes less than the minimal growth a_w after 100 h. There again, a stunted growth of *L. innocua* is observed.

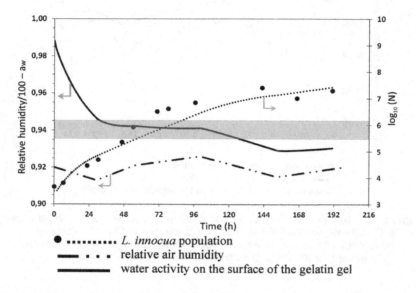

● ·········· *L. innocua* population
▬▬ · · · · relative air humidity
▬▬▬▬▬ water activity on the surface of the gelatin gel

Figure 4.7. *Growth of L. innocua on the surface of a gelatin gel during drying at 18°C, 4.6 m/s and 92.6% HR. The points correspond to the experimental measurements and the lines correspond to the predictions of the model except for the relative humidity of the air which is measured*

Thus, the thermodynamic model has allowed the completion of the information given by the mass transfer model and the understanding of the differences in behavior depending on the environmental conditions.

Conclusion

The Extensions of the Thermodynamic Approach

Physical-chemical properties play an important role at the level of the different qualities of biological media: nutritional, biochemical, mechanical, organoleptic or technological. To be able to predict their evolution according to the composition of the medium or according to the applied technological processing becomes a necessity in order to meet the challenges that industries must face using biological media such as pharmaceutical and food industries and in general biotechnology industries.

The proposed thermodynamic approach makes it possible to respond to the prediction of two of the Physical-chemical properties: the pH and the a_w. The works carried out in recent years have extended the area of applicability of the model, first to liquid, then to gelled and finally to solid biological products. The results obtained provide a means to quickly respond to questions such as: if ingredient X is replaced by ingredient Y, how does the pH and the a_w evolve? Similarly, its integration in a process simulator allows the completion of the information given by the heat and matter transfer models and provides the possibility of predicting the evolution of a microbial population, of the color of a product or of its mechanical qualities.

Additional works are underway in several directions: first, the prediction of pK_a of the molecules present in biological media or in food products, then the prediction of the redox potential in biological media; finally, the extension of the thermodynamic model toward sulfur compounds. To carry out these works, a number of tools from quantum mechanics are used and have enabled initial breakthroughs to be made. Therefore, the pK_a of organic acids or peptides could be calculated [TOU 13]. Similarly, for the redox potential, the first results refer to simple binary or ternary media. The redox potential of systems such as water + ascorbic acid is predicted with an accuracy of the order of 20 mV.

As a result, it is perfectly conceivable to have a complete tool for the prediction of the three significant Physical-chemical properties that are the pH, the water activity and the redox potential that can be integrated in a process simulator involving biological media or food products.

APPENDICES

Appendix 1

The UNIFAC Model

The selected model is a modified form of the UNIFAC equation proposed by Larsen *et al.* [LAR 87]. The activity coefficient of the compound i is the sum of two terms:

$$\ln(\gamma_i) = \ln(\gamma_i^C) + \ln(\gamma_i^R) \qquad [A1.1]$$

where:

γ_i^C is the combinatorial contribution of the activity coefficient of the compound i, and

γ_i^R is the residual part of the activity coefficient of the compound i.

A1.1. Combinatorial contribution

For a mixture of N components, the combinatorial contribution is written as:

$$\ln(\gamma_i^C) = \ln\left(\frac{\Phi_i}{x_i}\right) + 1 - \frac{\Phi_i}{x_i} \qquad [A1.2]$$

where γ_i^C is the combinatorial contribution of the activity coefficient of the compound i, Φ_i is molecular volume fraction of the compound i in the mixture and x_i is the molar fraction of the compound i.

The molecular volume fraction of the molecule i is calculated using the relation:

$$\Phi_i = \frac{x_i \cdot r_i^{2/3}}{\sum_{j=1}^{j=N} x_j \cdot r_j^{2/3}}$$ [A1.3]

where r_i is the molecular van der Waals volume of the molecule i and N is the number of compounds in the solution.

The parameter r_i is derived from the volume parameters of the groups constituting the compound i:

$$r_i = \sum_{k=1}^{k=NG} v_k^{(i)} \cdot R_k$$ [A1.4]

where R_k is the group volume of group k, $v_k^{(i)}$ is the number of groups k in the molecule i and NG is the number of groups.

$$R_k = \frac{V_{vdw}}{15.17}$$ [A1.5]

$$Q_k = \frac{A_{vdw}}{2.5 \cdot 10^9}$$ [A1.6]

where Q_k is the area group of group k, V_{vdw} is the Van der Waals volume, cm^3/mol and A_{vdw} is the Van der Waals surface area, $cm^2/mole$.

The values of Q_k and R_k used in the Achard model [ARC 92] are given in Appendix 4.

A1.2. Residual contribution

The expression of the residual contribution of the activity coefficient is given by:

$$ln(\gamma_i^R) = \sum_{k=1}^{k=NG} v_k^{(i)} \left[ln(\Gamma_k) - ln\left(\Gamma_k^{(i)}\right) \right]$$ [A1.7]

where Γ_k is the residual activity coefficient of the group k and $\Gamma_k^{(i)}$ is the residual activity coefficient of the group k in a reference solution containing only the compound i.

The residual part of the activity of a group k coefficient is written as:

$$ln(\Gamma_k) = \sum_{k=1}^{k=NG} Q_k$$

$$\left[1 - ln\left(\sum_{m=1}^{m=NG} \Theta_m \cdot \Psi_{mk}\right) - \sum_{m=1}^{m=NG} \frac{\Theta_m \cdot \Psi_{mk}}{\sum_{p=1}^{p=NG} \Theta_p \cdot \Psi_{pm}} \right] \qquad [A1.8]$$

where Θ_m is the volume fraction of group m in the solution and Ψ_{mk} is the interaction term between groups m and k.

The surface fraction of a group in a solution is expressed in a similar manner to the surface fraction of a compound in a solution:

$$\Theta_m = \frac{X_m \cdot Q_m}{\sum_{p=1}^{p=NG} X_p \cdot Q_p} \qquad [A1.9]$$

where X_m is the Mole fraction of group m in the solution.

$$X_m = \frac{\sum_{j=1}^{j=N} v_m^{(j)} x_j}{\sum_{p=1}^{p=NG} X_p \cdot Q_p} \qquad [A1.10]$$

The interaction term follows an Arrhenius law

$$\Psi_{mn} = exp\left(-\frac{a_{mn}}{RT}\right) = exp\left(-\frac{u_{mn} - u_{nn}}{RT}\right) \qquad [A1.11]$$

where a_{mn} is the binary interaction parameter between groups m and n, u_{mn} is the interaction energy of the UNIFAC groups.

$\Gamma_k^{(i)}$ is calculated with relations [A1.8] to [A1.11]: the designated sums with the indexes k, m and p refer only to existing groups in the reference solution composed of the molecule i.

The values of a_{mn} and a_{nm} used in the Achard model [ARC 92a] are given in Appendix 5.

Appendix 2

Debye–Hückel Expression for Long-Range Interactions

The extended form of the Debye–Hückel equation proposed by Pitzer [PIT 73a] takes account of electrostatic forces. The relation for the activity coefficient is given by:

$$\ln\left(\gamma_i^{LR}\right) = -\left(\frac{1000}{M_S}\right)^{0,5} A_\phi \left[\left(\frac{2z_i^2}{\rho}\right)\ln\left(1 + \rho I_x^{0,5}\right) + \frac{z_i^2 I_x^{0,5} - 2I_x^{1,5}}{1 + \rho I_x^{0,5}}\right] \quad [A2.1]$$

where M is the molecular mass of the solvent (g mol^{-1}), A_ϕ is the Debye–Hückel coefficient, z_i is the absolute value of the ionic charge, ρ is the parameter of the Pitzer expression and I_x is the ionic strength, mole fraction base.

The Debye–Hückel coefficient is equal to:

$$A_\phi = \frac{1}{3}\left(\frac{2.\pi.N_A.d_S}{1000}\right)^{0,5}\left(\frac{e^2}{\varepsilon_S.k.T}\right)^{1,5} \quad\quad [A2.2]$$

where NA is the Avogadro number (6.0225×10^{23} mole^{-1}), dS is the liquid, solvent density (g cm^{-3}), e is the electron charge (4.802654×10^{-10} e.u.), ε_S is the dielectric constant of the solvent i, and k is the Boltzmann constant (1.38048×10^{-23} m^2.kg.s^{-2}.K^{-1}) and T is the temperature (K).

The ionic strength is thus expressed as:

$$I_x = \frac{1}{2} \sum x_i z_i^2 \quad\quad [A2.3]$$

where x_i is the molar fraction of the compound i and z_i is the absolute value of the ionic charge.

Finally, the dielectric constant of the solvent is given by the following equation:

$$I_x = \frac{1}{2} \sum x_i z_i^2 \qquad \text{[A2.4]}$$

It should be noted that A_ϕ can be rewritten in the form $A_\phi = 1.40 \times 10^6 . \frac{\sqrt{d_s}}{(\varepsilon_S.T)^{1,5}}$ and is then expressed in $(g/mol)^{0.5}$.

Appendix 3

Solvation Equations

In what follows, the notation H always refers to the standard hydrated state. The variables corresponding to the conventional standard state have no index.

A3.1. Hydrated species structure parameters

$$R_k^H = R_k + N_{h_k} \cdot R_{water} \qquad\qquad [A3.1]$$

$$Q_k^H = Q_k + N_{h_k} \cdot Q_{water} \qquad\qquad [A3.2]$$

where R_k is the parameter relating to the volume of group k, Q_k is the parameter relating to the surface of group k and N_{h_k} is the hydration number of the ion k at infinite dilution.

A3.2. Mole fractions corresponding to the standard hydrated state

Let us take an N-constituent liquid mixture where constituent 1 represents water. The other N-1 species are solvated by water.

The water mole fraction is equal to:

$$x_k^H = \frac{x_1 - \sum_{j=2}^{j=N} N_{h_j} \cdot x_j}{1 - \sum_{j=2}^{j=N} N_{h_j} \cdot x_j} \qquad\qquad [A3.3]$$

where N_{h_j} is the hydration number of ion k at infinite dilution and x_j is the molar fraction of compound i.

The molar fraction of the ionic species is:

$$x_i^H = \frac{x_i}{1 - \sum_{j=2}^{j=N} N_{h_j} x_j}$$

[A3.4]

A3.3 Analytical relation between γ_i^{SR} and $\gamma_i^{SR,H}$

$$\gamma_1^{SR} = \gamma_1^{SR,H} \frac{x_1^H}{x_1}$$

[A3.5]

$$\text{for } i \neq 1 \ \gamma_i^{SR} = \gamma_i^{SR,H} \frac{x_i^H}{x_i} \left[\gamma_i^{SR,H} . x_i^H \right]^{-N_{h_i}}$$

[A3.6]

In practice, $\gamma_1^{SR,H}$ and $\gamma_i^{SR,H}$ are calculated with the modified UNIFAC equation (relations [A1.1]–[A1.11]) using the mole fractions and the structural paramaters of the hydrated species. Relations [A3.5] (water) and A [A3.6] (solutes) are then used to calculate γ_1^{SR} and γ_i^{SR}.

Formulas [A3.5] and [A3.6] satisfy the Gibbs–Duhem fundamental equation.

Appendix 4

List of Functional Groups

The values of the constants R_k and Q_k are those given by Larsen *et al.* [LAR 87] [groups 1–21] and by Achard [ARC 92] [groups 22–66].

Group			Subgroup		R_k	Q_k
Name	Number	Formula	Number	Formula		
Alkanes	1	CH_2	1	CH_3	0.9011	0.848
			2	CH_2	0.6744	0.540
			3	CH	0.4469	0.228
			4	C	0.2195	0.000
Alkenes	2	C=C	5	CH_2=CH	1.3454	1.176
			6	CH=CH	1.1168	0.867
			7	CH2=C	1.1173	0.985
			8	CH=C	0.8887	0.676
			9	C=C	0.6606	0.485
Aromatic	3	ACH	10	ACH	0.5313	0.400
			11	AC	0.3652	0.120
Alcohols	4	OH	12	OH	1.0000	1.200
	5	CH_3OH	13	CH_3OH	1.0000	1.000
Water	6	H2O	14	H2O	0.9200	1.400
Ketones	7	CH_2CO	15	CH_3-C=O	1.6724	1488
			16	CH_2-C=O	1.4457	1.488
Aldehydes	8	CHO	17	CHO	0.9980	0.948
Esters	9	CCOO	18	CH_3COO	1.9031	1.728
			19	CH_2COO	1.6764	1.420

Group			Subgroup		R_k	Q_k
Name	Number	Formula	Number	Formula		
Ethers	10	CH_2O	20	CH_3O	1.1450	0.900
			21	CH_2O	0.9183	0.780
			22	CHO	0.6908	0.650
			23	FCH_2O	0.9183	1.100
Amines	11	NH_2	24	NH_2	0.6948	1.150
	12	CH_2NH	25	CH_3NH	1.4337	1.050
			26	CH_2NH	1.2070	0.936
			27	CHNH	0.9795	0.624
	13	CH_2N	28	CH_3N	1.1865	0.940
			29	CH_2N	0.9597	0.632
	14	ANH_2	30	ANH_2	0.6948	1.400
Pyridines	15	Pyridine	31	C_5H_5N	2.9993	2.113
			32	C_5H_4N	2.8332	1.833
			33	C_5H_3N	2.6670	1.553
Nitriles	16	CH_2CN	34	CH_3CN	1.8701	1.724
			35	CH_2CN	1.6434	1.416
Acids	17	COOH	36	COOH	1.3013	1.224
CCl	18	CCl	37	CH_2Cl	1.4654	1.264
			38	CHCl	1.2380	0.952
			39	CCl	1.0060	0.724
CCl_2	19	CCl_2	40	CH2Cl2	2.2564	1.988
			41	CHCl2	2.0606	1.684
			42	CCl2	1.8016	1.448
CCl_3	20	CCl_3	43	CHCl3	2.8700	2.410
			44	CCl3	2.6401	2.184
CCl_4	21	CCl_4	45	CCl4	3.3900	2.910

Table A4.1. *Neutral groups*

Group		Subgroup		R_k	Q_k	Hydration
Number	Name					
22	H^+	46	H^+	0.4661	0.6018	2.959
23	K^+	47	K^+	0.3910	0.5350	2.957
24	Na^+	48	Na^+	0.1517	0.2847	2.606

Group		Subgroup		R_k	Q_k	Hydration
Number	Name					
25		49	NH_4^+	0.4862	0.6190	1.502
		50	NH_3^+	0.4290	0.4420	1.502
		51	NH_2^+	0.3100	0.2210	1.502
		52	NH^+	0.1900	0.0000	1.502
		53	N	0.0600	0.0000	1.502
26	Ag^+	54	Ag^+	0.3326	0.4806	1.013
27	Cs^+	55	Cs^+	0.7745	0.8442	3.395
28	Li^+	56	Li^+	0.0523	0.1399	2.532
29	Ca^{2+}	57	Ca^{2+}	0.1613	0.2967	3.077
30	Mg^{2+}	58	Mg^{2+}	0.0478	0.1320	3.928
31	Mn^{2+}	59	Mn^{2+}	0.0851	0.1937	0.85
32	Ba^{2+}	60	Ba^{2+}	0.4001	0.5435	0.41
33	Cd^{2+}	61	Cd^{2+}	0.1517	0.2848	1.292
34	Cu^{2+}	62	Cu^{2+}	0.0621	0.1569	1.517
35	Zn^{2+}	63	Zn^{2+}	0.0674	0.1658	2.487
36	Be^{2+}	64	Be^{2+}	0.0071	0.0370	6.25
37	Ni^{2+}	65	Ni^{2+}	0.0564	0.1441	2.715
38	Co^{2+}	66	Co^{2+}	0.0621	0.1569	3.296
39	Pb^{2+}	67	Pb^{2+}	0.2873	0.4359	1.408
40		68	Fe^{2+}	0.0674	0.1658	2.829
		69	Fe^{3+}	0.0460	0.1280	2.829
41	Sr^{2+}	70	Sr^{2+}	0.2336	0.3797	1.853
42	Ce^{3+}	71	Ce^{3+}	0.1838	0.3236	4.131
43	Nd^{3+}	72	Nd^{3+}	0.1638	0.2997	4.32
44	La^{3+}	73	La^{3+}	0.1744	0.3125	3.867
45	Cr^{3+}	74	Cr^{3+}	0.0416	0.1201	5.1
46	Pr^{3+}	75	Pr^{3+}	0.1728	0.3106	4.126
47	Sm^{3+}	76	Sm^{3+}	0.1490	0.2813	4.261
48	Al^{3+}	77	Al^{3+}	0.0221	0.0787	5.606
49	Sc^{3+}	78	Sc^{3+}	0.0652	0.1622	4.493
50	Y^{3+}	79	Y^{3+}	0.1184	0.2414	4.967
51	Eu^{3+}	80	Eu^{3+}	0.1426	0.2732	4.314
52	Rb^+	81	Rb^+	0.5282	0.6541	2.547
53	Oh^-	82	Oh^-	0.3912	0.5354	0

Group		Subgroup		R_k	Q_k	Hydration
Number	Name					
54	cl⁻	83	cl⁻	0.9860	0.9917	0
55	Br⁻	84	Br⁻	1.2520	1.1629	0
56	F⁻	85	F⁻	0.3912	0.5354	0
57	I⁻	86	I⁻	1.7706	1.4651	0.0145
58		87	NO_3^-	0.9537	0.9699	0.476
		88	NO_2^-	1.1770	1.1160	0.476
59	ClO_4^-	89	ClO_4^-	2.2987	1.7436	0.44
60	BrO_3^-	90	BrO_3^-	0.6073	0.7179	0.998
61	CNS⁻	91	CNS⁻	1.6069	1.3733	0.0745
62	ClO_3^-	92	ClO_3^-	0.8314	0.8851	0.804
63	CrO_4^{2-}	93	CrO_4^{2-}	2.7898	1.9838	0
64		94	SO_4^{2-}	2.8557	2.0149	0
		95	SO_3^{2-}	1.3300	1.2110	0
		96	Hso^{3-}	0.8170	0.8750	0
		97	HSO_4^-	1.1410	1.0930	0
		98	$S_2O_3^{2-}$	2.5980	1.8920	0
		99	$S_2O_4^{2-}$	2.9220	2.0460	0
65		100	HS⁻	1.4750	1.2970	0
		101	S^{2-}	1.0360	1.0250	0
		102	HCO_3^-	0.6310	0.7370	0
		103	CO_3^{2-}	0.9380	0.9590	0
		104	CN⁻	1.1950	1.1040	0
		105	$H_2PO_4^-$	1.3310	1.2110	0
		106	HPO_4^{2-}	1.7860	1.4630	0
		107	PO_4^{3-}	2.2420	1.7150	0
		108	-COO⁻	1.1750	0.7810	0
		109	HCOO⁻	2.4120	3.2620	0
66	neut	110	neut	1.0000	1.0000	0

Table A4.2. *Charged groups*

Appendix 5

Parameters of Interaction Between Functional Groups

		1	2	3	4	5	6	7
		CH$_2$	C=C	ACH	OH	CH$_3$OH	H$_2$O	CH$_2$CO
1	CH$_2$	0.000	76.460	62.880	972.800	1318.000	1857.000	414.000
		0.000	-0.183	-0.249	0.269	-0.013	-3.322	-0.516
		0.000	-0.366	1.108	8.773	9.000	-9.000	1.803
2	C=C	-46.450	0.000	35.070	633.500	1155.000	1049.000	577.500
		-0.182	0.000	-0.080	0.000	0.000	-3.305	0.000
		-0.489	0.000	0.376	0.000	0.000	0.000	0.000
3	ACH	-1.447	-0.028	0.000	712.600	979.800	1055.000	87.640
		-0.056	-0.071	0.000	-1.459	-1.793	-2.968	-0.462
		-1.612	-0.341	0.000	9.000	3.844	9.854	6.691
4	OH	637.500	794.700	587.300	0.000	29.500	155.600	161.000
		-5.832	0.000	-0.679	0.000	0.404	0.376	0.750
		-0.870	0.000	9.000	0.000	0.000	-9.000	9.000
5	CH$_3$OH	16.250	-6.808	10.970	66.340	0.000	-75.410	-29.400
		-0.300	0.000	-0.073	-0.585	0.000	-0.757	-0.729
		0.692	0.000	0.497	0.000	0.000	-4.745	-1.670
6	H$_2$O	410.700	564.400	736.700	-47.150	265.500	0.000	40.200
		2.868	0.000	1.965	-0.495	3.540	0.000	1.668
		9.000	0.000	0.000	8.650	8.421	0.000	-1.994
7	CH$_2$CO	71.930	-144.300	92.190	179.600	263.300	272.400	0.000
		-0.796	0.000	0.613	-1.285	-0.155	-1.842	0.000
		-2.916	0.000	-8.963	-4.007	1.768	0.330	0.000
8	CHO	313.500	161.800	125.400	2553.000	-274.000	0.000	-53.040
		-4.064	0.000	-3.133	0.000	0.000	0.000	-0.627
		0.000	0.000	0.000	0.000	0.000	0.000	0.000

		1	2	3	4	5	6	7
		CH_2	C=C	ACH	OH	CH_3OH	H_2O	CH_2CO
9	CCOO	44.430	200.300	8.346	266.900	394.000	245.000	43.650
		-0.972	0.000	-0.525	-1.054	-0.561	-0.072	0.191
		0.552	0.000	0.000	3.586	-0.101	2.754	0.000
10	CH_2O	369.900	-17.230	125.200	137.100	295.200	183.100	160.400
		-1.542	-1.648	-1.093	-1.115	-0.219	-2.507	0.548
		-3.223	0.000	0.590	-4.438	3.441	0.000	0.000
11	NH_2	346.500	454.900	902.700	-173.700	-297.400	-244.500	0.000
		1.595	0.000	-5.763	1.642	0.830	0.286	0.000
		0.000	0.000	0.000	0.000	0.000	0.000	0.000
12	CH_2NH	149.500	17.420	188.900	-233.900	-440.500	-342.400	0.000
		1.336	0.000	0.097	1.737	-0.083	2.640	0.000
		0.000	0.000	8.732	0.000	-2.128	13.090	0.000
13	CH_2N	-64.360	28.080	-95.460	-287.600	-440.200	-265.500	0.000
		-0.174	0.000	1.292	0.331	0.332	0.000	0.000
		1.135	0.000	0.000	-1.907	-2.960	0.000	0.000
14	ANH_2	680.500	0.000	334.300	170.300	170.100	498.300	-330.700
		-5.470	0.000	-1.655	0.000	1.738	0.000	-0.153
		0.000	0.000	0.000	0.000	0.000	0.000	0.000
15	Pyr	-52.030	0.000	-62.930	28.720	-283.600	-58.200	-48.010
		-0.555	0.000	-0.140	-0.257	0.626	1.231	0.000
		0.000	0.000	-0.970	9.000	0.000	1.509	0.000
16	CH_2CN	21.690	-64.530	-3.280	291.100	367.100	233.600	-249.400
		-1.226	0.000	-0.581	-0.276	-0.767	-1.448	-0.212
		0.000	0.000	-0.712	0.000	0.000	0.639	0.000
17	COOH	171.500	227.300	62.320	-92.210	757.200	86.440	-151.700
		-1.463	0.000	0.000	0.000	1.502	0.994	0.000
		0.676	0.000	0.000	0.000	0.000	-12.740	0.000
18	CCl	-67.330	340.300	-39.670	818.200	892.300	862.100	-93.460
		-0.679	0.000	-1.457	-4.270	-2.420	-2.637	0.678
		2.036	0.000	-0.078	-2.607	-1.140	0.000	0.000
19	CCl_2	12.870	48.610	240.700	716.600	947.500	856.500	409.500
		0.265	-1.484	-0.183	0.000	-3.570	-2.549	-0.382
		0.000	0.000	0.000	0.000	0.000	0.000	0.768
20	CCl_3	-35.460	38.700	210.500	708.600	1029.000	837.800	245.200
		-0.123	0.109	-0.296	-2.613	-4.307	0.000	-2.128
		1.134	0.117	0.000	7.771	-8.902	0.000	4.221
21	CCl_4	27.880	-95.030	80.230	918.500	1273.000	1323.000	378.700
		-0.166	-0.692	0.387	-2.045	-1.618	0.000	-0.467
		-0.609	0.000	1.830	9.000	9.000	0.000	-2.138

		8	9	10	11	12	13	14
		CHO	CCOO	CH$_2$O	NH$_2$	CH$_2$NH	CH$_2$N	ANH$_2$
1	CH$_2$	721.500	329.100	230.500	420.700	248.000	217.700	580.800
		-1.470	-0.152	-1.328	-2.256	-1.800	-0.151	-2.310
		0.000	-1.824	-2.476	0.000	0.972	1.117	-16.050
2	C=C	320.400	-24.650	321.600	243.800	223.900	54.210	0.000
		0.000	0.000	4.551	0.000	0.000	0.000	0.000
		0.000	0.000	0.000	0.000	0.000	0.000	0.000
3	ACH	215.100	97.300	82.860	72.600	29.250	88.030	307.600
		1.936	0.190	0.611	-0.430	-0.185	-1.130	-0.945
		0.000	-0.751	-0.739	0.000	-2.193	0.000	-8.767
4	OH	-325.200	169.100	227.000	-176.500	-199.900	196.800	-58.270
		0.000	0.190	1.364	-0.107	-0.475	3.925	0.000
		0.000	4.625	3.324	-1.016	0.000	0.000	0.000
5	CH$_3$OH	177.200	-49.460	-73.540	-182.900	-201.700	79.020	79.900
		0.000	-0.776	-1.237	1.257	3.930	2.560	-2.152
		0.000	0.469	-2.308	0.000	0.000	0.000	0.000
6	H$_2$O	0.000	218.000	19.540	-66.390	111.500	-15.800	-193.600
		0.000	-0.427	1.293	-1.053	-3.302	0.000	0.000
		0.000	-6.092	-8.850	0.000	9.347	0.000	0.000
7	CH$_2$CO	76.100	-11.930	-48.000	0.000	0.000	0.000	798.900
		0.920	-0.041	-0.510	0.000	0.000	0.000	0.380
		0.000	0.000	0.000	0.000	0.000	0.000	0.000
8	CHO	0.000	-133.600	220.400	0.000	0.000	0.000	0.000
		0.000	0.000	1.738	0.000	0.000	0.000	0.000
		0.000	0.000	0.000	0.000	0.000	0.000	0.000
9	CCOO	241.700	0.000	277.000	0.000	312.500	0.000	0.000
		0.000	0.000	0.326	0.000	0.000	0.000	0.000
		0.000	0.000	0.000	0.000	0.000	0.000	0.000
10	CH$_2$O	-17.530	-129.400	0.000	0.000	13.400	0.000	0.000
		-0.712	-0.041	0.000	0.000	-0.396	0.000	0.000
		0.000	0.000	0.000	0.000	0.000	0.000	0.000
11	NH$_2$	0.000	0.000	0.000	0.000	0.000	0.000	0.000
		0.000	0.000	0.000	0.000	0.000	0.000	0.000
		0.000	0.000	0.000	0.000	0.000	0.000	0.000
12	CH$_2$NH	0.000	-129.300	92.970	0.000	0.000	0.000	0.000
		0.000	0.000	0.000	0.000	0.000	0.000	0.000
		0.000	0.000	0.000	0.000	0.000	0.000	0.000

		8	9	10	11	12	13	14
		CHO	CCOO	CH₂O	NH₂	CH₂NH	CH₂N	ANH₂
		0.000	0.000	0.000	0.000	0.000	0.000	0.000
13	CH₂N	0.000	0.000	0.000	0.000	0.000	0.000	0.000
		0.000	0.000	0.000	0.000	0.000	0.000	0.000
		0.000	0.000	0.000	0.000	0.000	0.000	0.000
14	ANH₂	0.000	0.000	0.000	0.000	0.000	0.000	0.000
		0.000	0.000	0.000	0.000	0.000	0.000	0.000
		0.000	0.000	0.000	0.000	212.500	0.000	0.000
15	Pyr	0.000	0.000	0.000	0.000	0.105	0.000	0.000
		0.000	0.000	0.000	0.000	0.000	0.000	0.000
		0.000	-210.300	0.000	0.000	0.000	0.000	380.800
16	CH₂CN	0.000	0.167	0.000	0.000	0.000	0.000	0.000
		0.000	0.000	0.000	0.000	0.000	0.000	0.000
		0.000	-224.600	-248.100	0.000	0.000	0.000	0.000
17	COOH	0.000	-0.723	0.000	0.000	0.000	0.000	0.000
		0.000	0.000	0.000	0.000	0.000	0.000	0.000
		-15.230	0.000	154.600	-129.300	0.000	0.000	0.000
18	CCl	1.532	0.000	0.000	0.000	0.000	0.000	0.000
		0.000	0.000	0.000	0.000	0.000	0.000	0.000
		0.000	74.000	-58.450	0.000	0.000	-131.700	0.000
19	CCl₂	0.000	1.064	1.549	0.000	0.000	0.000	0.000
		0.000	0.000	0.000	0.000	0.000	0.000	0.000
		0.000	180.000	127.500	0.000	0.000	-77.130	0.000
20	CCl₃	0.000	0.519	2.681	0.000	0.000	6.007	0.000
		0.000	-0.797	0.000	0.000	0.000	0.000	0.000
		0.000	121.400	97.700	337.400	168.900	240.400	384.000
21	CCl₄	0.000	0.000	0.653	0.000	0.000	0.000	-0.032
		0.000	0.000	1.000	0.000	0.000	0.000	0.000

		15	16	17	18	19	20	21
		Pyr	CH$_2$CN	COOH	CCl	CCl$_2$	CCl$_3$	CCl$_4$
1	CH$_2$	273.800	559.000	664.100	264.300	101.200	103.100	-12.650
		0.176	0.454	1.317	0.258	-0.847	-0.124	0.045
		0.000	0.000	-4.904	-0.414	0.000	-1.818	0.336
2	C=C	0.000	294.400	186.000	-135.300	-15.950	-26.140	148.000
		0.000	0.000	0.000	0.000	1.216	0.124	0.924
		0.000	0.000	0.000	0.000	0.000	0.000	0.000
3	ACH	99.330	198.300	537.400	49.150	-195.100	-181.600	-57.830
		0.283	0.960	0.000	1.656	-0.141	0.114	-0.351
		1.530	1.398	0.000	-2.346	0.000	0.000	-1.284
4	OH	311.800	77.890	61.780	194.700	230.700	112.800	415.300
		2.405	-0.433	0.000	0.546	-3.591	1.955	1.391
		9.000	0.000	0.000	0.000	0.000	8.077	9.000
5	CH$_3$OH	491.800	-6.177	-321.200	-45.130	-88.720	-138.600	-36.750
		-2.773	-0.378	-1.116	-0.126	-0.018	0.433	-0.020
		0.000	0.000	0.000	0.105	-3.794	-0.291	0.840
6	H$_2$O	472.300	338.400	8.621	527.000	596.400	674.500	705.500
		1.336	1.900	-1.709	1.416	3.071	0.000	2.540
		1.756	-0.206	6.413	0.000	0.000	0.000	0.000
7	CH$_2$CO	170.000	387.500	230.000	194.400	-285.700	-252.000	-49.290
		0.000	0.044	0.000	-0.974	-0.015	1.399	0.436
		0.000	0.000	0.000	0.000	-1.678	-6.332	4.367
8	CHO	0.000	0.000	0.000	227.300	0.000	0.000	0.000
		0.000	0.000	0.000	-3.680	0.000	0.000	0.000
		0.000	0.000	0.000	0.000	0.000	0.000	0.000
9	CCOO	0.000	346.500	557.900	0.000	-142.800	-222.500	46.320
		0.000	0.113	1.377	0.000	-0.669	0.248	0.000
		0.000	0.000	0.000	0.000	-0.291	1.833	0.000
10	CH$_2$O	0.000	0.000	286.600	156.200	-64.570	-241.800	188.700
		0.000	0.000	0.000	0.000	-0.661	-0.431	-1.081
		0.000	0.000	0.000	0.000	0.000	-0.410	1.659
11	NH$_2$	0.000	0.000	0.000	571.800	0.000	0.000	42.360
		0.000	0.000	0.000	0.000	0.000	0.000	0.000
		0.000	0.000	0.000	0.000	0.000	0.000	0.000
12	CH$_2$NH	612.100	0.000	0.000	0.000	0.000	0.000	-14.720
		6.987	0.000	0.000	0.000	0.000	0.000	0.000
		0.000	0.000	0.000	0.000	0.000	0.000	0.000

		15	16	17	18	19	20	21
		Pyr	CH$_2$CN	COOH	CCl	CCl$_2$	CCl$_3$	CCl$_4$
13	CH$_2$N	0.000	0.000	0.000	0.000	-207.700	-403.900	-248.300
		0.000	0.000	0.000	0.000	0.000	-0.049	0.000
		0.000	0.000	0.000	0.000	0.000	0.000	0.000
14	ANH$_2$	0.000	-187.100	0.000	0.000	0.000	0.000	809.900
		0.000	0.000	0.000	0.000	0.000	0.000	-4.505
		0.000	0.000	0.000	0.000	0.000	0.000	0.000
15	Pyr	0.000	64.570	0.000	0.000	-272.000	-119.900	-162.700
		0.000	0.000	0.000	0.000	0.000	0.879	-0.021
		0.000	0.000	0.000	0.000	0.000	0.000	0.000
16	CH$_2$CN	23.650	0.000	0.000	0.000	0.000	-81.430	-61.770
		0.000	0.000	0.000	0.000	0.000	0.000	-0.432
		0.000	0.000	0.000	0.000	0.000	0.000	0.386
17	COOH	0.000	0.000	0.000	113.700	-73.880	56.260	148.300
		0.000	0.000	0.000	0.000	0.000	-1.041	-0.708
		0.000	0.000	0.000	0.000	0.000	0.000	0.000
18	CCl	0.000	0.000	447.800	0.000	-95.080	-106.600	42.600
		0.000	0.000	0.000	0.000	0.000	0.000	0.000
		0.000	0.000	0.000	0.000	0.000	0.000	0.000
19	CCl$_2$	367.500	0.000	617.300	108.100	0.000	0.000	-50.420
		0.000	0.000	0.000	0.000	0.000	0.000	-0.459
		0.000	0.000	0.000	0.000	0.000	0.000	0.000
20	CCl$_3$	-37.940	126.000	455.600	145.100	0.000	0.000	-24.810
		-0.276	0.797	1.316	0.000	0.000	0.000	-0.028
		0.000	0.000	0.000	0.000	0.000	0.000	0.065
21	CCl$_4$	323.600	495.200	500.300	35.100	141.900	47.170	0.000
		-0.032	0.770	1.533	-0.141	0.260	-0.058	0.000
		0.000	1.477	0.000	0.000	0.000	-0.130	0.000

Bibliography

[ABR 75] ABRAMS D.S., PRAUSNITZ J.M., "Statistical thermodynamics of liquid mixtures: a new expression for the excess Gibbs free energy of partly or completely miscible systems", *American Institute of Chemical Engineering Journal*, vol. 21, pp. 116–128, 1975.

[ACH 92a] ACHARD C., Modélisation des propriétés d'équilibre de milieux biologiques et alimentaires à l'aide de modèles prédictifs, thesis, University Blaise Pascal, Clermont-Ferrand, 1992.

[ACH 92b] ACHARD C., DUSSAP C.G., GROS J.-B., "Prédiction de l'activité de l'eau, des températures d'ébullition et de congélation de solutions aqueuses de sucres par un modèle UNIFAC", *I. A. A.*, vol. 109, pp. 93–101, 1992.

[ACH 94] ACHARD C., DUSSAP C.G., GROS J.-B., "Prediction of pH in complex aqueous mixtures using a group-contribution method", *AIChE Journal*, vol. 40, pp. 1210–1222, 1994.

[ALL 79] ALLANEAU B., Influence des lipides sur la rétention de constituants volatiles: aspects thermodynamiques, thesis, Ecole Nationale Supérieure des Industries Alimentaires, Massy, France, 1979.

[BAB 14] BABU U.S., GAINES D.M., WU Y. *et al.*, "Use of flow cytometry in an apoptosis assay to determine pH and temperature stability of shiga-like toxin 1", *Journal of Microbiological Methods*, vol. 75, no. 2, pp. 167–171, 2014.

[BAU 00] BAUCOUR P., DAUDIN J.D., "Development of a new method for fast measurement of water sorption isotherms in the high humidity range validation on gelatine gel", *Journal of Food Engineering*, vol. 44, pp. 97–107, 2000.

[BEN 10] BEN GAÏDA L., DUSSAP C.G., GROS J.B., "Activity coefficients of concentrated strong and weak electrolytes by a hydration equilibrium and group contribution model", *Fluid Phase Equilibria*, vol. 289, no. 1, pp. 40–48, 2010.

[BIM 02] BIMBENET J.J., DUQUENOY A., TRYSTRAM G., *Génie Des Procédés Alimentaires – Des Bases Aux Applications*, Technique et Ingénierie, Dunod, 2002.

[BOT 91] BOTTIN J., FOURNIE R., MALLET J.-C., *Cours de chimie, 2ème année*, Dunod, 1991.

[CAT 94] CATTÉ M., DUSSAP C.G., GROS J.-B., "Excess properties and solid-liquid equilibria for aqueous solutions of sugars using a UNIQUAC model", *Fluid Phase Equilibria*, vol. 96, pp. 33–50, 1994.

[CAT 95] CATTÉ M., DUSSAP C.G., GROS J.-B., "A physical chemical UNIFAC model for aqueous solutions of sugars", *Fluid Phase Equilibria*, vol. 105, pp. 1–25, 1995.

[CHE 82] CHEN C.-C., BRITT H.I., BOSTON J.F. *et al.*, "Local composition model Gibbs energy of electrolyte systems. Part 1. Single solvent, single completely dissociated electrolyte system", *AIChE Journal*, vol. 28, pp. 588–596, 1982.

[CHI 84] CHIRIFE J., RESNIK S.L., "Unsaturated solutions of sodium chloride as reference sources of water activity at various temperatures", *Journal of Food Science*, vol. 49, no. 6, pp. 1486–1488, 1984.

[COM 00] COMAPOSADA J., GOU P., PAKOWSKI Z. *et al.*, "Desorption isotherms for pork meat at different NaCl contents and temperatures", *Drying Technology*, vol. 18, no. 3, pp. 723–746, 2000.

[DEB 23] DEBYE P., HÜCKEL E., "The theory of electrolytes. I. Lowering of freezing point and related phenomena", *Physikalische Zeitschrift*, vol. 24, pp. 185–206, 1923.

[DEM 01] DE MENDOÇA A.J.G., VAZ M.I.P.M., DE MENDONÇA D.I.M.D., "Activity coefficients in the evaluation of food preservatives", *Innovative Food Science & Emerging Technologies*, vol. 2, no. 3, pp. 175–179, 2001.

[DES 04] DESNIER-LEBERT I., Prédiction de la croissance de Listeria innocua par une approche phénoménologique: modélisations complémentaires des propriétés du milieu, des transferts d'eau et des cinétiques, PhD Thesis, University Blaise Pascal, Clermont-Ferrand, France, 2004.

[DOE 82] DOE P.E., HASHMI R., POULTER R.G. *et al.*, "Isohalic sorption isotherms. I. Determination for dried salted cod", *Journal of Food Technology*, vol. 17, pp. 125–134, 1982.

[ECK 02] ECKERT F., KLAMT A., "Fast solvent screening via quantum chemistry: COSMO-RS approach", *AIChE Journal*, vol. 48, pp. 369–385, 2002.

[FER 82] FERRO FONTAN C., CHIRIFE J., SANCHO E. *et al.*, "Analysis of a model for water sorption phenomena in foods", *Journal of Food Science*, vol. 47, no. 5, pp. 1590–1594, 1982.

[FRE 75] FREDENSLUND A., JONES R.L., PRAUSNITZ J.M., "Group contribution estimation of activity coefficients in nonideal liquid mixtures", *AIChE Journal*, vol. 21, pp. 1086–1099, 1975.

[GME 82] GMELING J., RASMUSSEN P., FREDENSLUND A., "Vapor-liquid equilibria by UNIFAC group contribution. Revision and extension. 2", *Industrial & Engineering Chemistry Process Design and Development*, vol. 21, pp. 118–127, 1982.

[GME 88] GMEHLING J., "Present status of group-contribution methods for synthesis and design of chemical processes", *Fluid Phase Equilibria*, vol. 144, pp. 37–47, 1988.

[GME 98] GMEHLING J., LOHMAN J., JAKOB A. *et al.*, "A modified UNIFAC (Dormund) model. 3. Revision and extension", *Industrial & Engineering Chemistry Research*, vol. 37, pp. 4876–4882, 1998.

[GRO 03] GROS J.B., DUSSAP C.G., "Estimation of equilibrium properties in formulation or processing of liquid foods", *Food Chemistry*, vol. 82, pp. 41–49, 2003.

[GUI 02] GUINEE T.P., FEENEY E.P., AUTY M.A.E. *et al.*, "Effect of pH and calcium concentration on some textural and functional properties of mozzarella cheese", *Journal of Dairy Science*, vol. 85, pp. 1655–1669, 2002.

[HAL 68] HALA E., PICK J., FRIED V. *et al.*, *Vapor-Liquid Equilibrium*, Pergamon Press, London, p. 599, 1968.

[HAN 91] HANSEN H.K., RASMUSSEN P., FREDENSLUND A. *et al.*, "Vapor-liquid equilibria by UNIFAC group contribution. 5. Revision and extension", *Industrial & Engineering Chemistry Research*, vol. 30, pp. 2352–2355, 1991.

[HAR 14] HARKOUSS R., SAFA H., GATELLIER PH. *et al.*, "Building phenomenological models that relate proteolysis in pork muscles to temperature, water and salt content", *Food Chemistry*, vol. 151, pp. 7–14, 2014.

[IGL 77] IGLESIAS H.A., CHIRIFE J., "Effect of fat content on the water sorption isotherm of air dried minced beef", *Lebensmittel Wissenschaft und Technologie*, vol. 10, p. 151, 1977.

[KIK 91] KIKIC I., FERMEGLIA M., RASMUSSEN P., "UNIFAC prediction of vapour-liquid equilibria in mixed solvent-salt systems", *Chemical Engineering Science*, vol. 46, no. 11, pp. 2775–2780, 1991.

[KLA 95] KLAMT A., "Conductor-like screening models for real solutions: a new approach to the quantitative calculation of solvation phenomena", *Journal of Physical Chemistry*, vol. 99, p. 2224, 1995.

[KOJ 79] KOJIMA K., TOGICHI K., *Prediction of Vapour-Liquid Equilibria by ASOG Method*, Kodansha-Elsevier, New York, 1979.

[LAR 87] LARSEN B.L., RASMUSSEN P., FREDENSLUND A., "A modified UNIFAC group-contribution model for prediction of phase equilibria and heats of mixing", *Industrial & Engineering Chemistry Research*, vol. 26, pp. 2274–2286, 1987.

[LEB 84a] LEBERT A., RICHON D., "Infinite dilution activity coefficients of n-alcohols as a function of dextrin concentration in water-dextrin systems", *Journal of Agricultural and Food Chemistry*, vol. 32, pp. 1156–1161, 1984.

[LEB 84b] LEBERT A., RICHON D., "Study of the influence of solute (n-alcohols and n-alkanes) chain length on their retention by purified olive oil", *Journal of Food Science*, vol. 49, no. 5, pp. 1301–1304, 1984.

[LEB 04] LEBERT I., DUSSAP C.G., LEBERT A., "Effect of a_w, controlled by the addition of solutes or by water content, on the growth of *Listeria innocua* in broth and in a gelatine model", *International Journal of Food Microbiology*, vol. 94, pp. 67–78, 2004.

[LEB 05] LEBERT I., DUSSAP C.G., LEBERT A., "Combined physico-chemical and water transfer modelling to predict bacterial growth during food processes", *International Journal of Food Microbiology*, vol. 102, pp. 305–322, 2005.

[LEM 92] LE MAGUER M., "Thermodynamics and vapor–liquid equilibria", in SCHWARTZBERG H.G., MARTEL R.W. (eds), *Physical Chemistry of Foods*, Marcel Dekker, New York, pp. 1–45, 1992.

[LIN 99] LIN S.-T., SANDLER S.I., "Infinite dilution activity coefficients from Ab initio solvation calculations", *AIChE Journal*, vol. 45, pp. 2606–2618, 1999.

[LIN 02] LIN S.-T., SANDLER S.I., "A priori phase equilibrium prediction from a segment contribution solvation model", *Industrial & Engineering Chemistry Research*, vol. 41, pp. 899–913, 2002.

[LOH 01] LOHMANN J., JOH R., GMEHLING J., "From UNIFAC to modified UNIFAC (Dortmund)", *Industrial & Engineering Chemistry Research*, vol. 40, pp. 957–964, 2001.

[MUC 91] MUCK R.E., O'KIELY P., WILSON P.K., "Buffering capacities in permanent pasture grasses", *Irish Journal of Agricultural Research*, vol. 30, no. 2, pp. 129–141, 1991.

[NIN 99] NINNI L., CAMARGO M.S., MEIRELLES A.J.A., "Modelling and prediction of pH and water activity in aqueous amino acid solutions", *Computers and Chemical Engineering Supplement*, vol. 3, pp. S383–S386, 1999.

[NIN 00] NINNI L., CAMARGO M.S., MEIRELLES A.J.A., "Water activity in polyol systems", *Journal of Chemical and Engineering Data*, vol. 45, pp. 654–660, 2000.

[OH 14] OH J.-H., VINAY-LARA E., MCMINN R. *et al.*, "Evaluation of NaCl, pH, and lactic acid on the growth of Shiga toxin-producing Escherichia coli in a liquid Cheddar cheese", *Journal of Dairy Science*, vol. 97, no. 11, pp. 6671–6679, 2014.

[OUL 98] OULD MOULAYE C.B., Calcul des propriétés de formation en solution aqueuse des composés impliqués dans les procédés microbiologiques et alimentaires. Prédiction et réconciliation de données. Modélisation des équilibres chimiques et des équilibres entre phases, Thèse de l'Université Blaise Pascal-Clermont-Ferrand II, 1998.

[OZD 99] OZDEMIR M., SADIKOGLU H., "Use of UNIFAC in the prediction of water activity values of aqueous polyol solutions", *Journal of Food Processing and Preservation*, vol. 23, pp. 109–120, 1999.

[PAS 03] PASTORINO A.J., HANSEN C.L., MC-MAHON D.J., "Effect of pH on the chemical composition and structure-function relationships of Cheddar cheese", *Journal of Dairy Science*, vol. 86, pp. 2751–2760, 2003.

[PER 97] PERES A., MACEDO E.A., "A modified UNIFAC model for the calculation of thermodynamic properties of aqueous and nonaqueous solutions containing sugars", *Fluid Phase Equilibria*, vol. 139, pp. 47–74, 1997.

[PER 99] PERES A., MACEDO E.A., "Prediction of thermodynamic properties using a modified UNIFAC model: application to sugar industrial systems", *Fluid Phase Equilibria*, vol. 158–160, pp. 391–399, 1999.

[PIT 73a] PITZER K.S., "Thermodynamics of electrolytes. 1. Theoretical basis and general equation", *Journal of Physical Chemistry*, vol. 77, pp. 268–277, 1973.

[PIT 73b] PITZER K.S., MAYORGA G., "Thermodynamics of electrolytes. II. Activity and osmotic coefficients for strong electrolytes with one or both ions univalent", *The Journal of Physical Chemistry*, vol. 77, no. 19, pp. 2300–2308, 1973.

[PRA 99] PRAUSNITZ J.M., LICHTENTHALER R.N., DE AZEVEDO E.M., *Molecular Thermodynamics of Fluid-Phase Equilibria*, 3rd ed., Prentice Hall Inc., Upper Saddle River, New Jersey, 1999.

[RAH 97] RAHMAN S.M., PERERA C.O., "Evaluation of the GAB and Norrish models to predict the water sorption isotherms in foods", in JOWITT R. (ed.), *Engineering & Food at ICEF 7*, Sheffield Academic Press, Sheffield, UK, pp. A101–A104, 1997.

[RAH 07] RAHMAN S.M., *Handbook of Food Preservation*, 2nd ed., CRC Press, Boca Raton, Florida, 2007.

[RAH 09] RAHMAN S.M., *Food Properties Handbook*, 2nd ed., CRC Press, Boca Raton, Florida, 2009.

[REN 68] RENON H., PRAUSNITZ J.M., "Local compositions in thermodynamic excess functions for liquid-mixtures", *American Institute of Chemical Engineering Journal*, vol. 14, pp. 135–144, 1968.

[ROB 59] ROBINSON R.A., STOKES R.H., *Electrolytes Solutions*, 2nd ed., Butterworths, London, 1959.

[ROB 98] ROBLES-OLVERA V., BÉGOT C., LEBERT I. *et al.*, "An original device to measure bacterial growth on the surface of meat at relative air humidity close than 100%", *Journal of Food Engineering*, vol. 38, pp. 425–437, 1998.

[ROU 07] ROUGIER T., BONAZZI C., DAUDIN J.D., "Modeling incidence of lipid and sodium chloride contents on sorption curves of gelatin in the high humidity range", *LWT – Food Science and Technology*, vol. 40, pp. 1798–1807, 2007.

[SAN 97] SANCHO M.F., RAO M.A., DOWNING D.L., "Infinite dilution activity coefficients of apple juice aroma compounds", *Journal of Food Engineering*, vol. 34, pp. 145–158, 1997.

[SCO 56] SCOTT R.L., "Corresponding states treatment of nonelectrolyte solutions", *The Journal of Chemical Physics*, vol. 25, no. 2, pp. 193–207, 1956.

[SHE 09] SHEEHAN A., O'CUINN G., FITZGERALD R.J. *et al.*, "Distribution of microbial flora, intracellular enzymes and compositional indices throughout a 12 kg Cheddar cheese block during ripening", *International Dairy Journal*, vol. 19, pp. 321–329, 2009.

[SIM 02] SIMATOS D., "Propriétés de l'eau dans les produits alimentaires: activité de l'eau, diagrammes de phases et d'états", in LE MESTE M., LORIENT D., SIMATOS D. (eds), *L'eau dans les aliments*, Edition TEC and DOC, Paris, pp. 49–83, 2002.

[SKJ 79] SKJOLD-JOERGENSEN S., KOLBE B., GMEHLING J. *et al.*, "Vapor-liquid equilibria by UNIFAC group contribution. Revision and extension", *Industrial & Engineering Chemistry Process Design and Development*, vol. 18, no. 4, pp. 714–722, 1979.

[SPI 00] SPILIOTIS N., TASSIOS D., "A UNIFAC model for phase equilibrium calculations in aqueous and non aqueous sugar solutions", *Fluid Phase Equilibria*, vol. 173, pp. 39–55, 2000.

[TEN 81] TENG T.T., SEOW C.C., "A comparative study of methods for prediction of water activity of multicomponent aqueous solutions", *Journal of Food Technology*, vol. 16, pp. 409–419, 1981.

[TOG 90] TOGICHI K., TIEGS D., GMEHLING J. *et al.*, "Determination of new ASOG parameters", *Journal of Chemical Engineering of Japan*, vol. 23, pp. 453–463, 1990.

[TOU 13] TOURÉ O., DUSSAP C.-G., LEBERT A., "Comparison of predicted pKa values for some amino-acids, dipeptides and tripeptides, using Cosmo-RS, ChemAxon and ACD/labs methods", *Oil & Gas Science and Technology*, vol. 68, no. 2, pp. 178–197, 2013.

[TOU 14] TOURÉ O., Prédiction des propriétés d'équilibre dans les milieux biologiques et alimentaires par le modèle COSMO-RS, thesis, University Blaise Pascal, Clermont-Ferrand, 2014.

[TRU 03] TRUJILLO F.J., YEOW P.C., PHAM Q.T., "Moisture sorption isotherm of fresh lean beef and external beef fat", *Journal of Food Engineering*, vol. 60, pp. 357–366, 2013.

[WEI 87] WEIDLICH U., GMEHLING J., "A modified UNIFAC model. I – Prediction of VLE, h^E and γ^∞", *Industrial & Engineering Chemistry Research*, vol. 26, pp. 1372–1381, 1987.

[WIL 62] WILSON G.M., DEAL C.H., "Activity coefficients and molecular structure", *Industrial and Engineering Chemistry Fundamentals*, vol. 1, pp. 20–23, 1962.

[WIL 64] WILSON G.M., "A new expression for the excess free energy of mixing", *Journal of the American Chemical Society*, vol. 86, pp. 127–130, 1964.

[WIL 00] WILSON P.D.G., WILSON D.R., WASPE C.R., "Weak acids: dissociation in complex buffering systems and partitioning into oils", *Journal of the Science of Food and Agriculture*, vol. 80, no. 4, pp. 471–476, 2000.

Index

Printed in the United States
By Bookmasters